ECOLOGY

The Living World

Gerry Martino Stark
D. Ann Hill

KENDALL/HUNT PUBLISHING COMPANY
4050 Westmark Drive   Dubuque, Iowa 52002

All photos property of Gerry Martino Stark and D. Ann Hill unless otherwise noted.

Contents

Preface to the Student: A MUST TO READ

Ecology is the study of the environment. Included are all living and nonliving components and their relationships. What this means to you and me is that the earth and everything in it belongs to the study of ecology. How we relate to one another, to other organisms, and to the physical world is ecology. The term "ecology" comes from the Greek word for house, *oikos*. Our Earth is the "house," and we, as well as the other living organisms, are living on and in that house.

When we think about the term "ecology," we usually focus on air, water, and thermal pollution; oil spills; endangered and extinct species; and pesticides in our food. While this is of great importance, we still must remember to focus our attention on how all of the earth and its biotic (living) and abiotic (nonliving) components get together to form a "circle." This circle shows the ever-changing, developing relationships of living things with their environment. We are dependent on one another, not only for survival, but also for the enjoyment of life which we have been given.

We will study the flow, production, and usage of energy. The sun is the basis of our energy. It plays an amazing role which is basic and vital to our needs. The transfer of energy from the sun to organisms and from organisms to other organisms will be emphasized. What we do with and how we use this energy will dictate the quality of life for generations to come.

Placing yourself in the middle of an ever-changing relationship will give you an appreciation of every component's worth. Each is of value in the circle of life. Sit back. Look at a small piece of this circle—recycling; what is its importance? How does recycling affect your family? Your income? Your expenses? Surrounding areas? Industry? Other living organisms? Raw materials? Conservation? Wildlife? Water quality? Air quality? Government?

We are in the midst of "the recycling age." As the most able, living organism, we are responsible for asking and answering these pertinent questions. What is our role in recycling? How important is this role? How can we improve recycling? We have a responsibility to each other, to ourselves, and most of all to our Creator, to be good stewards of our resources. Understanding, taking the time to "stop and smell the roses," and taking the time to study our "living world" will bring an appreciation, an awareness, and pride to all of us. We are caretakers; let's learn how to do a good job.

You (God) put everything under his (man's) feet. All flocks and herds, and the beasts of the fields, the birds of the air, and the fish of the sea, all that swim the paths of the seas.

Psalm 8:6-8 (NIV Version)

—Gerry Martino Stark
—D. Ann Hill

Acknowledgments

We would like to give much credit and thanks to our families, students, colleagues, and editor who have given us a continued supply of support and encouragement. A special recognition goes to John Cunnyngham, our mentor, friend, and boss, who gave us the initiative to start this project. His faith in us supplied us with the motivation and determination to undertake and complete this endeavor.

In addition, there are also a number of people who have contributed in some way to the preparation of this manuscript. We thank each and every one of you: our Maker; our typist, Linda Marshall; our reviewer, John Cunnyngham; Kathy Polk; Mike Scanlon; Ed Hallman; Charlsey Ailes; Dick Stark; Jessica Stark; Jackie Stark; Teresa Newcomb; Gertrud McGreevy; Susan Antillon; Ruth Douffas; Aaron Hess; Russ Lathrope; Whittbold Nursery; Mary's Wicker Outlet; Publix; City Island Farmer's Market; Dick Stark Total Alignment and Brake; Hansard Service Center, Bush Gardens (Tampa); SeaWorld (Orlando); and all federal and state environmental agencies (noted in photo credits).

1

ENVIRONMENT

CLIMATE

OCEAN

LAND

ORGANISMS

GREENHOUSE EFFECT

YEARS

I. **ECOLOGY**

    A. Definition

    B. Differences between abiotic and biotic factors

    C. Limiting levels of factors

    D. Abiotic (physical) factors

        1. Sunlight (defined)
          a. food chain: producers + consumers
          b. photosynthesis
          c. other uses of solar energy

        2. Water (H_2O)
          a. precipitation
          b. homeostasis
          c. other uses of water

        3. Atmosphere (climate/weather)
          a. how atmosphere affects life
          b. greenhouse effect
          c. global warming
          d. CO_2 +fossil fuels
          e. maintaining the balance

        4. Soil (nutrients)
          a. elements in soil
          b. weathering
          c. ores
          d. soil particles
          e. "miniature world"

    E. Summary

1. Read each section for enjoyment.
2. Review study questions.
3. Go back through sections and find answers.
4. Write answers in spaces provided.
5. Check definitions if you need to in glossary in back of textbook.
6. Budget your time.
7. Follow the course in order. Don't "skip" around.

Things to Look for in This Chapter:

- Definitions of **all** highlighted words.
- Four abiotic factors.
- How a food chain functions.
- Uses of solar energy.
- Why water is so important.
- How the gases in the atmosphere maintain balance.
- How climate influences where we live.
- The role soil plays as an abiotic factor.
- The elements that are found in our soil.
  How minerals affect our life.

Definition

Ecology is the study of our environment, the interactions and relationships of all living and nonliving things. Knowledge of geology, biology, atmospheric studies, and geography becomes important in the understanding of our circle of life. The living (biotic) and non-living factors (abiotic) each contribute to numerous relationships.

Differences Between Abiotic and Biotic Factors

Abiotic factors are all nonliving, physical factors in our environment: sunlight, water, soil, and atmosphere (includes weather). **Biotic factors**, plants, fungi, animals, and man, depend upon the abiotic factors for their existence. The interaction between abiotic and biotic factors will be examined in this text.

One must understand that abiotic factors can limit the biotic factors in a particular environment. Abiotic factors provide a "limiting level," factors that regulate a population, upon these organisms. An example of a "limiting level" might be the available amount of direct sunlight (abiotic) on tomato plants (biotic) at Martino's Farm (an area). The amount of sunlight limits tomato plant growth in this area. Another example might be the amount of fresh water (abiotic) needed for Jessica's and Jackie's five fish (biotic) aquarium (an area).

Figure 1.1: ENVIRONMENTS ACROSS OUR NATION. Studying ecology develops an awareness and an appreciation of our living world. Stop! Look! Listen!

✔ THINGS TO KNOW AND DO:

Define:

* ecology

* abiotic

* biotic

Give at least two examples of abiotic and biotic factors.

abiotic

1.
2.

biotic

1.
2.

Create your own example of a "limiting level."

What areas of knowledge are needed in studying ecology? (Name 3.)

a.
b.
c.

➔ MODIFIED TRUE - FALSE: correct underlined word if necessary.

_______ 1. Abiotic factors can limit <u>biotic</u> factors in our environment.

_______ 2. Abiotic factors are also called <u>chemical</u> factors.

_______ 3. <u>Abiotic</u> factors include animals, plants, and mankind.

_______ 4. The amount of <u>sunlight</u> per day affects the growth of certain plants.

_______ 5. Abiotic factors and biotic factors <u>interact</u> to form "our living world."

✍ FILL IN THE BLANKS:

1. ________________________ factors provide a "limiting level."

2. The ____________________, ____________________, ____________________, and ____________________ are all considered abiotic factors.

3. The study of our environment is ____________________.

4. "Living" factors are called ____________________.

5. A knowledge of ____________________ would be helpful in the study of ecology.

✎ IN SEARCH OF...

The authors say "take time to stop and smell the roses." Why? On what page is the statement found?

____

____

____

Figure 1.2: ANIMALS IN A ZOO. Food, water, living space—all are limiting factors for these elephants living at Busch Gardens, Florida.

Abiotic Factors

In this chapter, we will discuss the four (4) physical factors (abiotic): sunlight, water, atmosphere, and soil.

Sunlight (A form of radiant energy)

The sun, shining down upon the earth, provides the **radiant** (light) energy required for the existence of life. All of our food-producing organisms, from the land and from the ocean, depend upon the sun's energy for their existence. The sun is the beginning point of any **food chain.** A food chain is the transfer of energy from a **producer** (green plants) to a **consumer** (insect, animal, and man).

The process by which the producer (green plants) produces food (carbohydrates) is called **photosynthesis**. Photosynthesis is the process by which sunlight, carbon dioxide (CO_2), minerals, and water (H_2O) form carbohydrates (food) in green plants. The sun's energy is collected by the **chlorophyll** molecule in green plants during photosynthesis.

On land, our producers in the food chain begin with green grass and plants (corn plant

Figure 1.3: **PAHALI COAST, HAWAII.** Can you identify three abiotic factors in this photo?

in Figure 1.4) In the ocean, the producer is algae (seaweed) as seen in Figures 1.5 and 1.6. The sun's availability is necessary to both types of producers. There are parts of our world where sunlight is abundant, as in the

Figure 1.4: CYCLE OF LIFE. The sun's energy and nutrients are absorbed by producers (corn), then transformed through the process of photosynthesis into food for consumers (mice). Other consumers (owls) feed on mice. Their remains decompose, returning nutrients to the soil for plant use. Each organism has its part to play in the cycle of life.

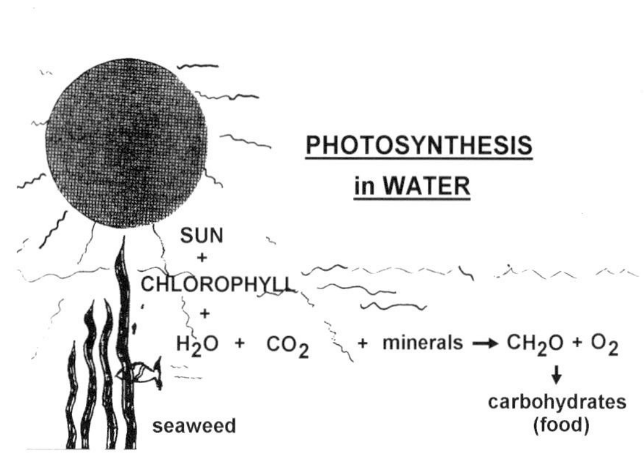

Figure 1.5A: PHOTOSYNTHESIS IN AN AQUATIC ENVIRONMENT. Seaweed (algae) is the producer, producing food for the ocean inhabitants.

Figure 1.5B: GAS EXCHANGE. Man exhales CO_2 which is taken in by plants. Plants give off O_2 which is then taken in by man. (A convenient exchange!)

desert and in tropical areas. In contrast, sunlight is limited on the floor of forests and in the depths of our oceans.

Considering the great amount of energy given off by the sun, only about 1/2 or 50% of solar radiation actually reaches the earth. This energy is mainly absorbed by land and water; only 1% of this solar radiation is used in photosynthesis; the other 49% that is absorbed by land and water is radiated back into the atmosphere. The remaining 50% of solar radiation never reaches Earth.

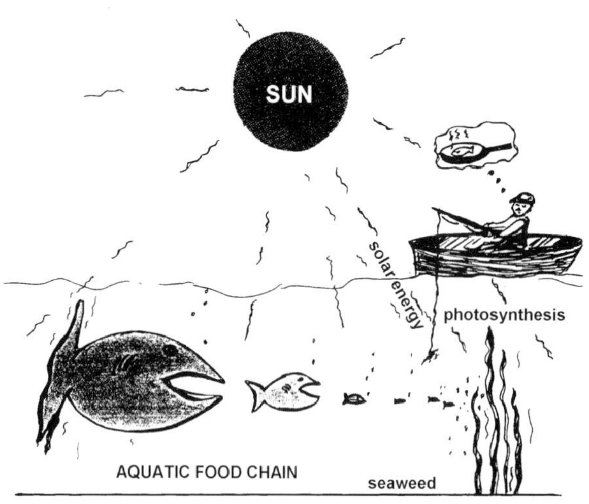

Figure 1.6: AQUATIC FOOD CHAIN. Photosynthesis in a marine environment begins with algae. What is man's role?

Figure 1.7A: SOLAR PANELS. Solar panels collect the sun's energy which is used as an electrical source for this car wash.

In addition to aiding in food production, solar radiation has the ability to provide energy for many other segments of our world. The abiotic factor (the sun) is collected by solar panels, stored and used to provide electricity for heating and, yes, for cooling our homes. Storage of collected solar radiation is becoming less and less a problem today. As technologies improve, efficiency will also increase. Future development of other uses of solar radiation is inevitable. Solar energy is free, clean, safe, and available. Best yet, it is non-polluting. Its abundance is continual; however, man must maximize the use of this energy source.

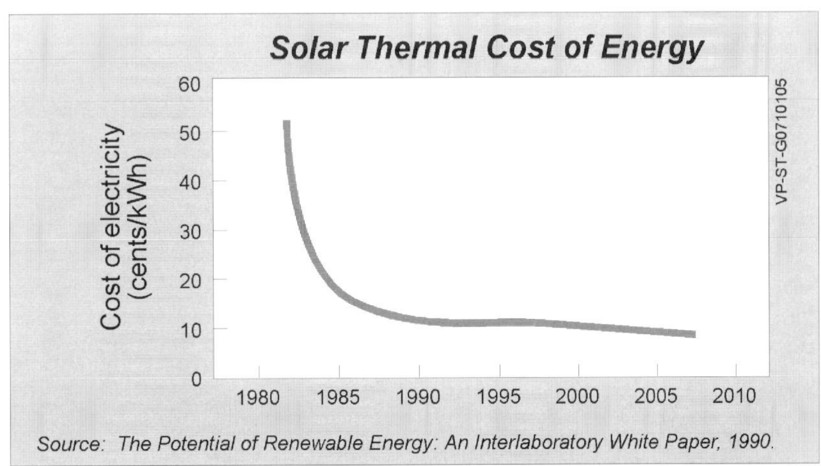

Figure 1.7B: COST OF SOLAR ENERGY. With the use of solar thermal power plants, the cost of electricity decreased by 83% in less than a decade. (U.S. Dept. of Energy)

✔ THINGS TO KNOW AND DO:

What is sunlight?____

Define:

* food chain ____

____

* producer ____

____

* consumer ____

____

Create your own example of a food chain. Include at least 3 organisms. Label energy, source, producer, and consumer.

Explain the process of photosynthesis.

____

____

____

Give the photosynthesis formula.

____

Show the relationship of gas exchange between plants and man. ____________________

____

____

What percentage of the sun's radiant energy is used in the process of photosynthesis? ______ %

Does all of the sun's energy reach the earth? (Give a percentage, if necessary.)

____

List 3 benefits of solar energy.

1. ____

2. ____

3. ____

What is the abiotic factor you learned in this section?

____

____

_________ 1.   Without the <u>sun</u>, life as we know it would not exist on the Earth.

____

_________ 2.   Chlorophyll <u>stops</u> the sun's energy. _______________________

_________ 3. CO_2 represents carbon <u>monoxide</u>. _______________________

_________ 4.   The food chain <u>begins</u> with the sun. _______________________

_________ 5.   Man is a <u>consumer</u>. _______________________

_________ 6.   In Figure 1.4 the mouse in the grave represents <u>decomposition</u>.

____

_________ 7.   Only <u>20%</u> of solar radiation reaches Earth. _______________________

_________ 8.   Man and plants take in the <u>same</u> gas. _______________________

_________ 9.   Solar energy is <u>limited</u> everywhere. _______________________

_________ 10. Photosynthesis is a process for making <u>proteins</u>. _______________________

_________ 1.   Radiant energy is _______________________ .

_________ 2.   Sunlight is plentiful in _______________________ and in _______________________ .

_________ 3.   Limited sunlight can be found in the _______________________ on the

floor of the _______________________ .

_________ 4.   The _______________________ is an abiotic factor.

_________ 5.   We use collected solar energy for _______________________ and

_______________________ our homes.

What does the grave stone represent?

Define the above term.

____

____

____

Who is Nick?

____

Water (H₂O)

About two-thirds of the Earth's surface is covered by H_2O. Life as we know it cannot exist without water (abiotic factor). Water comes from the atmosphere as **precipitation** to the earth as rain, sleet, snow or hail. It may fall into our oceans, lakes, streams, rivers and ponds, or may fall onto the land, seeping into the soil and into the underground water system.

Organisms may either drink or absorb water. In all organisms, some H_2O becomes chemically incorporated into their cells, as in photosynthesis. Water is absorbed by plant cells and is used in the production of more plant cells.

As you can see, water is necessary for all living organisms. An organism's internal activities on a cellular level are involved with the H_2O molecule. Water is necessary in maintaining a steady state or balance (**homeostasis**) within living things. Water balance within organisms as small as a **paramecium**, a single-celled organism, is extremely important. An organism such as the paramecium living in a watery environment depends on the proper amount of H_2O at all times. Too much water within this organism causes **cytolysis,** a bursting of the cell. Too much H_2O leaving the cell cause **plasmolysis** or shrinking of the cell. In

both cases, it leads to death of the organism. We also need water for both structure and function of each cell in our bodies. Besides regulating the function of biotic factors, water also regulates the temperature (abiotic factor) of our world.

Our oceans act somewhat as stabilizers on the climate of the earth. In the summer the waters absorb heat slowly, cooling the surrounding area. In the winter the ocean loses heat slowly, acting as a radiator giving up heat to the surrounding areas.

Figure 1.8: WATER IRRIGATION. Sprinklers bring water to dry land, a costly method. (U.S. Soil Conservation Service)

↘ DID YOU KNOW?

Figure 1.9: WINTER/SUMMER BEACH. During the summer, the water temperature is cooler than the air temperature, and in the winter the water temperature is warmer than the air temperature. In the winter and summer beach scenes, which photo demonstrates a "radiator" and which, an "air conditioner?"

✔ THINGS TO KNOW AND DO:

Define:

* homeostasis ____

____

____

* paramecium ____

____

____

Why is water necessary for life?

____

____

____

How is a water molecule used in a living organism?

____

____

Relate cytolysis and plasmolysis to a paramecium.

____

____

How much of the Earth's surface is covered by water? ____________________

Water is a (an) ____________________ factor.

What is a general term for all water falling to the earth's surface? ____________________

In what forms does the Earth obtain water?

____

____

How does the ocean control the Earth's temperature?

____

____

➡ MODIFIED TRUE-FALSE: correct underlined word if necessary.

__________ 1. Two-thirds of the Earth's surface is covered by water. ____________________

__________ 2. Water is not necessary for all living organisms.____________________

__________ 3. We acquire water from the atmosphere.____________________

__________ 4. Water can regulate biotic as well as abiotic factors.____________________

__________ 5. Water balance within a cell is vital.____________________

__________ 6. Cytolysis is caused by too little water.____________________

__________ 7. The temperature of the ocean helps heat the surrounding lands in the summer.
____

__________ 8. Water is used in the reproduction of plant cells.____________________

✎ FILL IN THE BLANKS:

1. In the _________________, waters absorb heat slowly, _________________the surrounding area.

2. _________________is the balance that is maintained within an organism.

3. _________________is a single-celled organism.

4. Sleet is a form of _________________.

5. H_2O is _________________.

6. Water is (a,an) _________________factor.

✎ IN SEARCH OF...

In this section what abiotic factor acts as a stabilizer? How?

____

____

Atmosphere (Weather/Climate)

Most of the Earth's atmosphere is only five to seven miles above its surface. This atmosphere is a reservoir of several gases (abiotic): nitrogen (78%), oxygen (21%), and carbon dioxide and water vapor (less than 1%). Each gas has its value. We living organisms need the oxygen from the atmosphere in order to live. Oxygen is inhaled during respiration and is used by our bodies in a variety of ways.

The most abundant gas in our atmosphere is **nitrogen**. This gas is used by **nitrogen-fixing bacteria** living in the roots of special plants (legumes). This bacteria takes nitrogen from the air, and through a series of reactions, "fixes" it into a more usable form for the plant's benefit. CO_2, another gas in the atmosphere, is used by green plants in the process of photosynthesis. This process produces carbohydrates, our food source.

In addition to providing the oxygen to breathe and the gases to aid in food production, the atmosphere helps screen out most of the dangerous **ultraviolet (UV)** rays of the sun. Within the past years, we have been warned about the ill-effects of UV radiation on our skin. The developing holes in the ozone layer are reducing the efficiency of the atmosphere to screen out the harmful UV rays. Skin cancer is on the upswing due to the increased UV rays reaching the Earth's surface. We all need to use sunscreen with at least a sun protection factor (SPF) of 15 before spending time outdoors.

If you are a fair-skinned person, you have less pigmentation in your skin, and you will burn easily. It is a proven fact that fair-haired, blue-eyed people have a higher risk of skin cancer. Since the ozone layer is protecting us less, even dark-skinned sunlovers must cover up with sunscreen.

Figure 1.10: ATMOSPHERE/CLIMATE, IMPORTANT ABIOTIC FACTORS. Along with the atmosphere, our climate is important. Temperature variations, amounts of precipitation, and amounts of sunlight have a great effect on our environment and the organisms that live in it.

Our atmosphere also has a role in regulating the temperature of the Earth. The atmosphere absorbs heat from the sun and holds this heat for a time close to the Earth's surface. Carbon dioxide and water vapor trap this heat, increasing the temperature of the atmosphere; this condition is known as the **"greenhouse effect."** Did you ever wonder why people say not to leave a dog in a closed car? The "greenhouse effect" comes into mind. Picture this: a sunny

Figure 1.11: SUNSCREEN. Two science students apply sunscreen for protection against the hot Florida noonday sun.

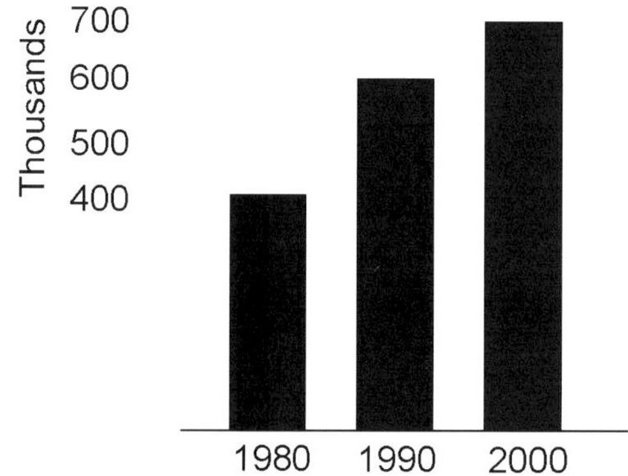

Figure 1.12: SKIN CANCER CASES. Studies indicate that the number of skin cancer cases are steadily rising. This increase is due to the developing holes in the ozone layer, allowing more harmful ultraviolet rays to reach Earth's surface.

Figure 1.13: "HOT DOG!" Remember not to leave a pet in a closed car on a hot, sunny day. As your pet exhales CO_2, the CO_2 traps heat in the car for a longer period of time, increasing the air temperature.

day, a closed car, your pet inside inhaling and exhaling the available air (atmosphere.) This sunlight, radiating through the windows, is transferred into heat energy. As your pet exhales CO_2, this added CO_2 traps the heat for a long period of time, thereby increasing the temperature of the air inside the car. In this mini-atmosphere within your car, temperature increases quickly; your dog is then overcome by the heat which may prove fatal.

Included in this example, you can see the **Law of Conservation of Energy** being demonstrated. This law states: "energy can neither be created nor destroyed but can be changed from one form to another," like the light energy being changed to heat energy in the car's atmosphere.

Now let's go back to the earth's atmosphere to see how an increase of CO_2 affects our environment. All living organisms respire in one form or another, releasing CO_2. The exhaling from respiration (CO_2) and the burning of fossil fuels (CO_2) increases the already existing level of CO_2. The more **fossil fuels** (oil, coal) we use and burn, the more CO_2 is released into the atmosphere. This increasing amount of CO_2 traps the heat and creates a problem. Forest fires and cooking fires release CO_2 as well. This added level of CO_2 in the atmosphere contributes to the rising temperature and the **"greenhouse effect"** on the Earth.

In a typical greenhouse, light energy from the sun enters in through the windows easily. This light energy is transferred into heat energy which remains for a long period of time. Remember this same transference happened with our pet in the closed car. This heat energy raises the temperature which provides a warmer climate for the plants in the greenhouse even during very cold outside temperatures. This provides a good environment for the plants as long as there is ample sunlight.

However, in our environment, the additional CO_2 in the atmosphere is not a benefit; it is a problem. This increasing of the temperature of the earth may cause **global warming**. Reducing the extra amount of CO_2 entering the atmosphere is the necessary starting point in solving the problem. Finding alternate energy sources to replace fossil fuels is one way to approach this problem. Solar energy may be used more frequently for a variety of energy uses like heating and cooling homes and generating electricity. In addition, we need to create devices which would cut down or reduce

Figure 1.14: GREENHOUSE. A typical greenhouse depends upon the "greenhouse effect" for additional heat during the winter season.

Figure 1.15: Recent studies have shown that global warming has caused a rise in temperature of 36° since the 1930s. This temperature increase has caused cracks in glaciers. What would the melting of a glacier do to coastline areas? (U.S. Geological Survey No. 425)

CO_2 from entering our atmosphere when fossil fuels are used. Finally, we might develop new uses which would be beneficial for the current levels of CO_2 being released.

Each of us may contribute our part to this solution. Let's cut down on our own use of fossil fuels. This is not a new problem. We have been told in recent years about the "greenhouse effect" and "global warming." Many studies have been done and are still being investigated. How could you and your friends help?

____

____

____

In the past our Earth has been able to absorb heat and radiate that heat back beyond our atmosphere with good efficiency. A delicate balance of absorbing and radiating heat back into space had been achieved. Now, because of the added amount of fossil fuels we are burning and CO_2 being generated from other areas, we are getting out of balance. CO_2 levels are increasing; heat is trapped longer and therefore, temperatures are rising. The ability to radiate this excess heat is decreasing. Good

research is being done at the present time to alleviate this developing situation, but we can't "hold our breath" (CO_2) until the problem is solved. Be concerned about what you can do; don't leave it up to someone else. Remember, we are all to be good stewards.

Along with the atmosphere, our **climate** (abiotic factor) is important. Temperature variations, amounts of precipitation, and amounts of sunlight have a great effect on our environment. Many organisms are temperature dependent; they can only exist within a certain temperature range. Other organisms depend upon a certain amount of rainfall and/or sunlight; consequently, the type and amount of these organisms is determined by the climate in an area.

The **climate** of an area depends upon the type of weather commonly found in a specific place. The humidity, precipitation, and temperature of an area is measured daily over a long period of time in order to accumulate data. The climate of an area makes specific

Figure 1.16: OLD OIL TANK. The majority of homes were heated by the burning of oil (fossil fuels) in the past. Oil was stored in tanks such as this. Today less polluting methods are being used to heat our homes, such as gas and solar energy.

Figure 1.17: WHAT'S WRONG WITH THESE PICTURES? Every organism must fit its environment or it will perish.

demands on the **flora** (plant life) and the **fauna** (animal life) in an area.

Not all types of flora and fauna can survive and flourish under all climate situations. Each organism has specific moisture and temperature requirements. Too much or too little moisture may cause death to an organism. An organism's temperature range must also be considered. Too much or too little heat may wipe out an entire species. **Extinction** (a dying out) of that species may follow if no suitable climate can be found.

There are many climates on Earth. We have hot, rainy areas like the tropical rain forests; hot and dry areas as in the desert and very cold areas like in the tundra. Areas where there are seasonal changes ranging from wet to dry and from hot to cold are found in **temperate** regions of the world. One temperate region includes the continental United States.

In conclusion, climate has a **limiting effect.** Changing weather conditions may cause plant and animal life to either survive and flourish or to diminish and die out.

"There is a time for everything, and a season for every activity under heaven."

(Ecclesiastes 3:1 NIV version of the *Bible*)

Figure 1.18: VARIOUS CLIMATES. Environments, such as these, are very different due to climatic conditions that exist within an area. Left desert photo is from a museum exhibit at Busch Gardens, Florida. Right photo is a rain forest. (U.S. Park Service)

↘ DID YOU KNOW?

In Cameroon in 1986, Lay Nyos released a large cloud of carbon dioxide, killing 1700 people. The CO_2 seeped into the lake, causing the surface waters to become totally saturated. An unknown disturbance caused the release of the gas; it pushed the oxygen aside and caused the people in the area to suffocate from the lack of oxygen, not from CO_2, which is non-poisonous.

✔ THINGS TO KNOW AND DO:

Where is the Earth's atmosphere located?

____

____

List the gases that make up our atmosphere. How much of each gas is found in the atmosphere (%)?

1. _______________________________________

2. _______________________________________

3. _______________________________________

How do we use these gases?

____

____

____

Name 3 ways that the atmosphere benefits us.

1. _______________________________________

2. _______________________________________

3. _______________________________________

What are UV rays? Explain.

____

____

____

What is SPF 15? Explain.

____

____

____

Describe the "greenhouse effect."

____

____

____

How does additional CO_2 interfere with the Earth's atmosphere?

____

____

____

What might happen to a dog in a closed car on a hot, sunny day?

____

____

____

Discuss the Law of Conservation of Energy.

____

____

____

____

What are fossil fuels? (Name 2.)

____

____

____

____

What is the term used to describe the rising temperature of the Earth?

____

Name one alternate source of energy that might replace fossil fuels.

____

____

Name one way you can personally help reduce the CO_2 level of our atmosphere.

____

____

What abiotic factor did you study in this section?

What does "temperature dependent" mean?

What is the difference between climate and weather?

How can climate affect an entire species?

Describe the climate in a tropical rain forest.

Describe the climate in the desert.

What is the importance of nitrogen-fixing bacteria? Where are they found?

Define flora and fauna and give an example of each.

Flora: _______________________________

Defined: _______________________________

Fauna: _______________________________

Defined: _______________________________

→ MODIFIED TRUE-FALSE: correct underlined word if necessary.

________ 1. The Earth's atmosphere is <u>10</u> miles above its surface. __________________

________ 2. The majority of our atmosphere is made up of <u>oxygen</u>. __________________

________ 3. UV rays come from the <u>sun</u>. __________________

________ 4. Fair-skinned people burn easily in the sun because they have <u>more</u> pigmentation in their skin. __________________

________ 5. Nitrogen and carbon dioxide are used by <u>animals</u> in food production.

____

________ 6. SPF stands for <u>skin pigmentation factor</u>. __________________

________ 7. Ultraviolet rays are <u>dangerous</u>. __________________

________ 8. The greenhouse effect is a <u>hoax</u>. __________________

✍ FILL IN THE BLANKS:

1. A dog in a closed car on a hot day is called __________________.

2. Fair-skinned, blue-eyed people have a __________________ risk of

__________________ __________________.

3. __________________ __________________ rays are harmful rays coming from the sun.

4. __________________ and __________________ are fossil fuels.

5. The __________________ and __________________ in an area are affected by climate.

6. The dying out of a species is called __________________.

7. A temperate region can be found __________________.

✎ IN SEARCH OF...

What event in this section demonstrates The Law of Conservation of Energy? Explain.

Soil (Nutrients)

The soil (**abiotic**) of Earth is a complex array of elements and decaying matter. The type and amount of elements found in the soil will determine the type of living organisms existing in a particular area.

Soil is formed by **weathering** of rock material over a long period of time; soil is also formed by the decomposition of **organic** (living) matter. Water, air, and living organisms are also found in soil.

Not all of the elements found in rock and soil are useful to all living things. Silicon, for example, is one of the most abundant elements but is used only by **diatoms**, microscopic one-celled organisms found in the ocean. Silicon and other elements are usually represented by their chemical symbols (Si). The following elements, calcium (Ca), phosphorus (P), potassium (K), sulfur (S), sodium (Na), chlorine (Cl), magnesium (Mg), iron (Fe), and manganese (Mn) are also found in the Earth's **crust**, the thin, outer layer of the Earth's surface. Calcium, for example, is an element used in building skeletons of certain organisms (coral and shells.) Many of the elements found in the soil of the Earth's crust are mined for their economic value. These elements are called "**ores.**" These are minerals necessary in the production of many useful items including cars, building materials, glass production, and other products. Soil is also used to provide minerals which act as nutrients for plant life.

Homes and shelters are also provided by the soil for animal life.

Hot and cold temperatures as well as wet and dry times help soil to be formed. This process of **weathering** breaks down rock material into smaller particles from which soil is formed. Large particles are called **sand**; medium-sized

Figure 1.19: MINING CAMP. A typical mining camp depicts the process of collecting our natural resources. (U.S. Geological Survey, Ransome, F.L. 526)

Figure 1.20: SOIL EROSION: "A HOLE IN ONE." Formed from wind erosion, this mountain-top boulder shows the tremendous damage wind can cause. (U.S. Geological Survey, Segerstrom, K. 555)

particles are called **silt**, and very fine particles are called **clay**. Each of these rock particles are surrounded with moisture. Air spaces are also found between each soil particle. The moisture and air in the soil are necessary for life activities in this "miniature" world. Organisms living in this world (soil) obtain the necessary air and water which are the requirements for their life.

Summary

The soil is an interrelationship between organic organisms and inorganic matter. There are many interactions going on between living and nonliving components of the soil. The soil is a miniature world of biotic and abiotic factors.

We must be careful to prevent our soil from being eroded, polluted, and wasted. We will consider soil, its importance, and its conservation in future chapters.

✔ THINGS TO KNOW AND DO

How is soil formed?

____

____

List everything that is found in soil (all elements, water, etc.).

____

____

For what is silicon (Si) used?

____

____

What is the relationship of coral and shells to calcium?

____

____

____

What is "weathering?"

____

____

____

What are mined for their economic value?

____

____

____

What does soil provide for plant life?

____

____

____

What does soil provide for animal life?

____

____

____

List the three names for the various particles of soil.

1. ____

2. ____

3. ____

How does soil impact your daily life? (We don't mean muddy shoes.)

____

____

____

➜ MODIFIED TRUE - FALSE: correct the underlined word if necessary.

________ 1.   Soil is made up of <u>both</u> organic and inorganic matter. ________________

________ 2.   Silicon is one of the most <u>scarce</u> elements. ________________

________ 3.   <u>Chlorine</u> is used in building coral and shells. ________________

________ 4.   <u>All</u> minerals are mined for their economic value. ________________

________ 5.   Silt is the <u>finest</u> particle of soil. ________________

________ 6.   Air in the soil is <u>necessary</u> for life activity. ________________

✍ FILL IN THE BLANKS:

1. ______________________, ______________________, ______________________, and
______________________ are found in the Earth's crust.

2. The most abundant element in the Earth's crust is ______________________.

3. ______________________ are microscopic organisms found in the ocean.

4. ______________________ breaks down rock material into smaller particles.

✎ IN SEARCH OF...

Explain what we mean by "the soil is a miniature world."

____

____

____

____

____

2

ENZYMES
CANOPY
OPEN SYSTEM
STABLE
YUCCA TREE
SUNLIGHT
TERRESTRIAL
ENERGY
MINERALS

II. **ECOSYSTEM:**

   A. Ecosystem defined

      1. Aquatic ecosystem

      2. Terrestrial ecosystem

   B. Producers: First component of an ecosystem

      1. Autotrophs defined

      2. Photosynthesis
         a. its limits
         b. its efficiency

   C. Consumers: Second component of an ecosystem

      1. Heterotrophs defined

      2. Decomposition
         a. its Rate
         b. its Importance
         c. its Organisms
         d. its Process
         e. its Pathways
         f. its Variables

   D. Abiotic Variables: Third component of an ecosystem

      1. Variables defined

      2. Example of abiotic variable

   E. Summary

When trying to memorize names in a list, arrange the list so the beginning letters of each word form a word. Example: The 3 parts of an ecosystem include

- plants,
- animals, and
- environment

P, A, E, doesn't spell anything so rearrange your list to spell P, E, A.
plants
environment
animals

Things to Look for in This Chapter:

- Definitions of **all** highlighted words.
- Three components of an ecosystem.
- The importance of producers/autotrophs.
- The importance of consumers/heterotrophs/decomposers.
- The importance of photosynthesis.
- The process of decomposition.
- Variables that influence decomposition.
- Examples of abiotic variables.

ECOSYSTEM

An **ecosystem** is a **community** of plants and animals in their environment. These three components, plants, animals, and environment, are linked together by energy cycles. One of these energy cycles involves solar energy being changed to chemical energy (food) by photosynthesis. An insect feeds on plant material obtaining nutrients and energy. As the insect completes its life cycle, it dies and its body decomposes. Decomposition releases nutrients back into the soil helping the next generation of plants to grow. To help you understand the cycling process of energy and nutrients, let's look at one ecosystem, "the lily pad."

In Figure 2.1, focus only on the lily pad. In this ecosystem sunlight and chlorophyll are involved in the process of photosynthesis. This process provides food for the organisms in this ecosystem. The input of the needed nutrients found in the water is necessary for the lily's growth. Also, insects receive energy from the lily pad during their feeding process. Later,

Figure 2.1: ECOSYSTEM. Lily Pad Pond. What abiotic and biotic factors do you see?

Figure 2.2: ECOSYSTEMS. Left is a tropical rain forest (Hawaii). Notice the various communities in this ecosystem. Right photo is Lake Okeechobee, Florida, a fresh water ecosystem and its surrounding plant life.

through decay of the lily pad and the insects, the nutrients are released once again into the **"open ecosystem."** So we see there is input into this system of energy (sun) and nutrients (found in water). The output of energy and nutrients is found in the recycling of the decayed plant and insect matter. This flowing of energy and nutrients into and out of the ecosystem is known as an "open system."

Once again we are concerned with biotic and abiotic factors, but now just in one small area like the lily pad ecosystem. All of the factors, both biotic and abiotic, that affect the lily pad ecosystem are relatively **stable.** This means the proper amounts of water, gas, nutrients, temperature, sunlight, animals, insects, and other organisms are all working together in balance (dynamic equilibrium.)

At first glance, the outward appearance of these factors seems fixed (stable); however, these same factors are constantly changing. There is a continual movement of energy and materials within this ecosystem. This fixed, moving balance is called **stable dynamic equilibrium.** Without this balance, the productivity and maintenance of any ecosystem would be destroyed.

Figure 2.3: LILY PAD ECOSYSTEM. Notice chlorophyll in two different populations used for the process of photosynthesis. All the other requirements needed for photosynthesis are present: H_2O, CO_2, nutrients, and sunlight.

Figure 2.4: BOUNDARIES OF AN ECOSYSTEM. The boundaries of an ecosystem may be gradual or definite. The lake ecosystem shows a definite boundary; however, the grass ecosystem graduates into a forest ecosystem.

An **ecosystem** is basically an energy-processing system. The components of the ecosystem have been there for a long period of time. The boundaries of the ecosystem depend on the ability of organisms to survive in that particular area under given environmental conditions. Within an ecosystem one factor or organism is influenced by another factor or organism. For example, the amount of sunlight will influence the amount of plant life able to grow in an area. The plant life has a direct effect on the amount of consumers living in that ecosystem, and so on.

Figure 2.5: A MEANDERING STREAM. Notice the surrounding landscapes. These terrestrial ecosystems are influenced by the stream.

Looking at different ecosystems, we can see that any one ecosystem exerts some influence on the surrounding ecosystems. Take a quick glance at Figure 2.5: A meandering stream.

Can you describe how the stream's ecosystem may affect the surrounding landscape and the organisms in the soil? How would the soil type affect the stream's waterflow? The aquatic organisms? The nutrients found in the water?

Any aquatic ecosystem is influenced by the terrestrial ecosystem that it flows through. The makeup of the sides and the bed of the stream influence the water. In addition, the flowing stream influences the land around and below it; therefore, a sandy river bed would influence the stream quite differently than a muddy stream bed would. The rate of waterflow, as well as the depth of water, may influence the types of organisms living in that area.

In a similar way, island ecosystems are influenced by their distance from continental land masses. Nutrients found in the runoff from the continent, seed disbursal, and debris in the air also exert their influence on the island ecosystem (see Figure 2.7).

After looking at these various ecosystems, you should know that an ecosystem is not independent from its surroundings. Recall the lily pad ecosystem. There, nutrients, energy, and gases were cycled around and around. The input and output of materials of this "open" system came from surrounding ecosystems. This gives new meaning to the phrase "no man is an island." Neither men, nor ecosystems, can exist without being influenced by their environment.

Figure 2.6: ESTUARY IN CENTRAL FLORIDA. This bottle provides a home for barnacles and various aquatic organisms. How would the change of tides affect the organisms living here and the river bed?

Figure 2.7: AN ISLAND ECOSYSTEM. Nutrients from the mainland wash up on Manana Island off the coast of Oahu, Hawaii. (R. Naedele)

➘ DID YOU KNOW?

NO MAN IS AN ISLAND

Figure 2.8: NO MAN EXISTS ALONE. We depend upon many things for our existence. Use this picture to answer the questions in the margin.

What are this man's problems?

1. _______________________________

2. _______________________________

3. _______________________________

4. _______________________________

✔ THINGS TO KNOW AND DO:

Define:

*Ecosystem _______________________________

____

____

*"open" system _______________________________

____

____

*cycling of nutrients _______________________________

____

____

*community _______________________________

____

____

Give examples of each of the following in an ecosystem:

biotic factors

a.____

b.____

abiotic factors

a.____

b.____

c.____

d.____

Name the three components of an ecosystem:

____

____

____

➡ MODIFIED TRUE - FALSE: correct the underlined word if necessary.

________ 1.   Energy and nutrients go into and out of a "closed" system.

   ____

________ 2.   An ecosystem is a combination of organisms in their underlined{chemical} environment.

   ____

________ 3.   All organisms living in one area constitute a community. ____________________

________ 4.   A flowing stream has influence on the land around it. ____________________

________ 5.   Island ecosystems are not influenced by continental land masses.

   ____

✍ FILL IN THE BLANKS:

1.   ____________________ and ____________________ are involved in the process of photosynthesis.

2.   ____________________ are released through the process of decomposition.

3.   Ecosystems are relatively ____________________.

4.   This fixed, moving balance is called ____________________ ____________________ ____________________.

5.   An aquatic ecosystem is influenced by the ____________________ ecosystem.

| ✎ **IN SEARCH OF...** |

Make a list of items you would find in a pond ecosystem which would be influenced by the surrounding ecosystems. (Hint: land types, nutrients, gases, etc)

1.____

2.____

3.____

PRODUCERS: AUTOTROPHS

In simplest terms, all ecosystems, aquatic and terrestrial, consist of 3 basic components: producers (plants), consumers (animals) , and abiotic matter (environmental factors).

Now let's look at each of the three components in detail. The first component we'll discuss is the producer. **Producers** are largely green plants and are **autotrophs**. Autotrophs are organisms that make their own food through photosynthesis and are commonly called **self-feeders**. These organisms capture the sun's energy, then, during photosynthesis, produce food from simple organic and inorganic substances.

These autotrophs are usually found in the upper layers of the ecosystem where most of the sunlight reaches. Two good examples would be the forest canopy (top layer of trees) and surface waters of oceans, lakes, and ponds.

Photosynthesis

As part of the producer component, we should look at photosynthesis in more detail. We'll discuss its limits and its efficiency.

You should recall that the energy of the sun is captured by chlorophyll (green pigmented molecule.) The process of photosynthesis converts CO_2 and H_2O into carbohydrates (food.)

Figure 2.9: AUTOTROPHS. Top photo shows pond surface covered with aquatic producers (algae); forest canopy (bottom photo) also is an area of high producer activity.

The carbohydrate that is produced in photosynthesis is **glucose**. Glucose, a simple sugar, is only one of the many compounds made from this process. The final products are complex sugars, free amino acids, proteins, fats, and fatty acids, vitamins, pigments, and co-enzymes. All of these products are thought to be synthesized (made) in the **chloroplast**, the structure that contains chlorophyll. These final products are found in different parts of the plant and under different environmental conditions. Mature leaves, for example, may produce just simple sugar (glucose). Young shoots and rapidly developing leaves may produce fats, proteins, and other products.

Figure 2.10: AQUATIC AND TERRESTRIAL PRODUCERS. Producers may live in different areas and appear very different, yet they serve the same function in any ecosystem.

✔ THINGS TO KNOW AND DO:

Define:

*producers ____

____

____

*glucose ____

____

____

*self-feeders ____

____

____

*canopy ____

____

____

*photosynthesis ____

____

____

Give an example of:

autotrophs ____

canopy design ____

Create an example of an ecosystem, listing the autotrophs that live there.

____

____

____

____

➜ **MODIFIED TRUE-FALSE: correct underlined word if necessary.**

________ 1. Producers are also <u>autotrophs.</u> ________________

________ 2. During photosynthesis <u>inorganic</u> substances are made. ________________

________ 3. Autotrophs are usually found in the <u>middle</u> layer of an ecosystem.
____

________ 4. The surface of a pond has a <u>high</u> rate of photosynthesis.
____

✍ **FILL IN THE BLANKS:**

1. Chlorophyll is found in the ________________ of plants.

2. The final products of photosynthesis are ________________, ________________,
________________, ________________, ________________,
________________, ________________, and ________________.

3. The energy of the ________________ is captured by the ________________.

4. Autotrophs are often called ________________.

✎ IN SEARCH OF...

What do fats, proteins, and glucose have to do with fruit trees?

Photosynthesis: Its Limits

The process of photosynthesis is limited by both environmental conditions and the type of community structure encountered. These limits include such environmental factors as light intensity, temperature, moisture, atmospheric gases, and soil nutrients. Several plant factors also limit the rate of photosynthesis: canopy design, photosynthetic capacity, and leaf area.

To understand the importance of leaf area on the rate of photosynthesis, we must consider many factors: Are the leaves in full sunlight? Are they facing the sun directly or at an angle? Are the leaves shaded by other leaves? The answers to these questions are important in determining the rate of photosynthesis. The adaptation of leaf position is important in obtaining maximum photosynthetic rate. Corn, grasses, beets, and turnips are plants that have

Figure 2.11: CONSUMERS. We depend upon plants as a major food source. Corn is an example of a producer with good leaf position, ensuring high photosynthetic activity.

achieved good leaf position and leaf area. The rate of photosynthesis is high for these types of plants.

Canopy design is also considered to have a limiting effect of the rate of photosynthesis. A **canopy** is a covering that is created by the tallest trees in an area. Some canopies are more dense than others, letting very little sunlight reach the forest floor. Other canopies vary in their design. This design is created by having different heights of trees, clusters of various tree heights, various amounts of trees per unit of area, and various types of trees.

Photosynthetic capacity is the ability of a plant to achieve a specific amount of photosynthetic activity. This depends on the number of leaves on the plant and the size of the leaves. Remember that most of photosynthesis takes place in the leaves of green plants. From this

Figure 2.12: GOOD LEAF POSITION. Shape, size, and position of leaves are important for maximum crop production. Notice the difference of leaf position in bottom photo. (U.S. Soil Conservation Service)

information, which of the following producers has a higher photosynthetic capacity, a cactus or a magnolia tree? Why?

____

____

____

____

Photosynthesis: Its Efficiency

In terms of energy input, photosynthesis is a rather inefficient process. It has been calculated that photosynthesis has an efficiency rate of about 38%. These calculations are usually done on a yearly, or growing season, basis. To state this in another way, only about 1% of the light that falls on a plant (**annual incident radiation**) is converted to chemical energy. Even the most productive areas of vegetation utilize no more than 3% of sunlight, yet the total amount of energy in the light that reaches the surface of this planet is enormous. Some 120 billion metric tons of new organic material is produced every year, despite the low conversion efficiency in photosynthesis.

Figure 2.13: RATE OF PHOTOSYNTHESIS IN VARIOUS PLANTS. A magnolia tree has many large, leathery leaves whereas the cactus and other desert plants have very small leathery leaves. (U.S. Park Service)

Figure 2.14: PRODUCERS WE DEPEND UPON. Fresh fruits and vegetables are some of the foods (producers) that we consumers need for our nutrition.

There are additional statistics which quote the efficiency rates for specific plants, but scientists feel these statistics are not to be accepted with a great deal of confidence. Temperature, moisture, nutrients in the soil, and other factors were not held constant in different plants that were tested. This caused the statistics to be suspect.

In conclusion, producers in an ecosystem are one of the necessary components. The producers use the process of photosynthesis to produce food for the ecosystem.

✔ THINGS TO KNOW AND DO:

List environmental factors that limit photosynthesis:

a.____

b.____

c.____

d.____

e.____

List plant factors that limit the rate of photosynthesis:

a.____

b.____

c.____

Pick a plant and discuss leaf area and position.

____

____

____

____

Would your choice of a plant in the previous question have a high rate of photosynthesis? Why or why not?

____

____

____

____

____

➜ MODIFIED TRUE-FALSE: correct underlined word if necessary.

__________ 1.   All canopies are created <u>equally</u>. __________________

__________ 2.   Leaf position is <u>important.</u> __________________

__________ 3.   Grasses have <u>good</u> leaf position. __________________

__________ 4.   Photosynthesis is limited by both environmental conditions and <u>population</u> structure. __________________

__________ 5.   Photosynthetic capacity is the ability of a plant to achieve <u>varied</u> amounts of photosynthetic activity. __________________

✎ FILL IN THE BLANKS:

1. Magnolia trees have a ______________________ photosynthetic capacity.

2. Canopies allow ______________________ sunlight to reach forest floors.

3. ______________________ is an inefficient energy process.

4. ______________________ metric tons of new organic material is produced every year.

5. Photosynthesis has an efficiency rate of about ______________________ per cent.

✎ IN SEARCH OF...

Why were the scientists in this chapter not confident with certain statistics?

____

____

____

____

____

____

CONSUMERS: HETEROTROPHS

The second component in any ecosystem is the consumer. Consumers are **heterotrophs**, organisms which cannot make their own food. Most organisms, with the exception of plants, are heterotrophs. They use the food produced by the autotrophs (plants) for their own nutrition.

Heterotrophs depend directly or indirectly on the energy stored in the tissues of the producers. In general, they take the food, rearrange it chemically, use it, and, in the end, **decompose** it into inorganic compounds once again.

These heterotrophs are composed of two subsystems, consumers and decomposers. These consumers and decomposers regulate the rate of energy flow and nutrient cycling. They also stabilize the ecosystem.

Figure 2.15: HETEROTROPHS. Feeding at night at a river bed, these heterotrophs ultimately depend upon autotrophs.

Decomposition

Consumers usually feed on living tissue. For example, a man consuming an apple or an ear of corn is acting as a consumer. **Decomposers**, on the other hand, break down matter into inorganic substances.

For example, bacteria or fungi living on leather, fallen leaves, or other matter acts as decomposers. Both consumers (man) and decomposers (fungi, bacteria) are heterotrophic organisms.

Decomposition: Its Rate

The most intensive activity of an heterotrophic organism occurs where there is a dense accumulation of organic matter. The upper layer of soil where leaves, grass, branches, insects, and other consumers lie dead on the forest floor is example of an area of high heterotrophic activity. In an aquatic environment, the sediment layers on the bottom of oceans and lakes is another area of high heterotrophic activity.

In both these ecosystems, the decomposition rate is high. Decomposition is the opposite of photosynthesis, in many aspects. De-

Figure 2.16: DECOMPOSERS. Estimate the number of shown producers and decomposers in this photo. Decomposers depend upon producers to meet their nutritional requirements. (K. Polk)

composition is the eventual breakdown of organic matter into its basic parts: carbon dioxide, water, inorganic nutrients, and the release of energy. Decomposition, as you can see, involves many organisms in the breaking down of tissue. Photosynthesis, on the other hand, is involved in the building up of organic matter.

Figure 2.17: DECOMPOSITION. Left photo shows compost heap; right photo shows the forest floor. Decomposition is taking place in both places, releasing nutrients and replacing them into the soil.

Decomposition: Its Importance

Without decomposition, life on Earth could not exist. The raw materials used by living organisms would be bound permanently in living tissue and would never be reused. Decomposition allows these valuable and needed materials to be used and reused time and time again. Consider this: if every leaf that ever fell to the ground was never broken down through the process of decomposition, we would be "eyeball deep" in leaves. But this picture is just a fraction of a percent of the amount of matter that would be covering our land (lithosphere) and filling our waters (hydrosphere). So we say, "Hats off to the fungi and bacteria," or in other words, "Thank God there's a fungus among us!"

Figure 2.18: IMPORTANCE OF DECOMPOSITION. The leaves in this picture represent a tiny fraction of the total amount which fall from this tree annually. Decomposition breaks down bound material, enriching the soil, releasing nutrients, and reducing the amount of leaf litter.

✔ THINGS TO KNOW AND DO:

Define:

*heterotrophs ____

____

____

*consumers ____

____

____

*decomposers ____

____

____

*lithosphere ____

____

____

*hydrosphere ____

____

____

Give two examples of decomposers.

a.____

b.____

Discuss the importance of decomposition of matter.

____

____

____

____

____

➡ MODIFIED TRUE-FALSE: correct underlined word if necessary.

__________ 1.   Consumers are <u>autotrophs.</u> __________________

__________ 2.   Heterotrophs are <u>stabilizers</u> in the ecosystem. __________________

__________ 3.   Decomposers <u>build up</u> matter. __________________

__________ 4.   The <u>lower</u> layers of soil have high heterotrophic activity. __________________

__________ 5.   Decomposition is the opposite of <u>respiration.</u> __________________

✎ FILL IN THE BLANKS:

1.   ____________________ and ____________________ have high heterotrophic activity.

2.   Decomposition is the eventual breakdown of ____________________.

3.   ____________________, ____________________, ____________________, and ____________________ are the end products of decomposition.

4.   ____________________ is the building up of organic matter.

5.   ____________________ ____________________ used by living organisms would be permanently bound if there were no decomposers.

✏ IN SEARCH OF...

Why should we be glad that there's a "fungus among us?"

____

____

____

____

✔ THINGS TO KNOW AND DO:

Define:

*saprophytes ____

____

*mechanical action ____

____

*succession ____

____

*cellulose ____

____

*microscopic ____

____

List six macroorganisms of decomposition in a terrestrial ecosystem.

a.____

b.____

c.____

d.____

e.____

f.____

List three macroorganisms of decomposition in an aquatic ecosystem.

a.____

b.____

c.____

Discuss the "succession" of decomposers in any ecosystem.

____

➜ MODIFIED TRUE-FALSE: correct underlined word if necessary.

________ 1.   Both microscopic and macroscopic fungi account for <u>25%</u> of the decomposers of soil. __________________

________ 2.   Bacteria are usually the major decomposers of <u>plant</u> matter.

____

________ 3.   Fungi are usually the major decomposers of <u>plant</u> matter. __________________

________ 4.   Nutrients fixed in the tissues of organisms are <u>available</u> for recycling.

____

________ 5.   Sugar-consuming fungi and bacteria invade the <u>roots</u> of plants.

____

✍ FILL IN THE BLANKS:

1.   ______________________ are macroscopic fungi.

2.   ______________________ % of decomposers are bacteria.

3.   ______________________ is the end point of a food chain.

4.   ___________________ and ___________________ are specialists in taking chemical energy out of cellulose.

5.   Microbes make ______________________ available for plant use.

✎ IN SEARCH OF...

Why would the authors say that decomposition is the ending of one food chain and the beginning of a new food chain? Explain.

____

____

____

____

____

Figure 2.24: HERBIVORE. Rabbit (consumer) is feeding on plant matter (producers). This is the early part of a food chain.

Figure 2.25: CARNIVORE. A snake waits for unsuspecting prey (rabbit in first photo). This is another step in the food chain.

Decomposition: Its Pathways

Decomposition reduces organic matter into inorganic matter. Remember that decomposition is not only the **end** of a food chain, but also the beginning of a **new** food chain. Now let's see how this process is accomplished.

One pathway of decomposition begins as **herbivores**, plant-eating organisms like rabbits and insects, feed. The **carnivores**, meat-eating organisms like foxes, lions, and hawks, consume their **prey**. The victims of carnivores are usually smaller carnivores or herbivores. These carnivorous animals extract minerals and nutrients from the food they eat for their own nutrition. They also **deposit** a substantial portion of partially decomposed material, **feces**, (solid waste products). This deposited fecal matter is then worked upon by fungi and bacteria.

On the other hand, decomposition of leaves actually begins while the leaves are still on the plant or tree. As the leaves approach **senescence**, old age, the plant itself reabsorbs much of the nutrients of the leaves into its permanent parts (stems, roots).

During the growing season, plants give off tiny quantities of material through their pores. These tiny droplets of material support **microflora**, microscopic plant life that **inhabits** (lives on) plants. Other microscopic organisms, take in organic material from the roots of living plants for their nutrition. The areas around the roots and the root surface support a variety of microbes. They feed on material that is given off by the root: simple sugars, fatty acids and amino acids.

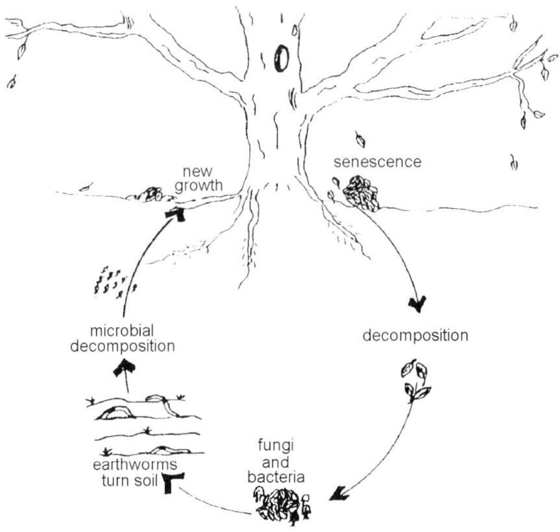

Figure 2.26: LEAF LITTER. The falling of leaves (senescence) allows the decomposers in and on the soil to release bound nutrients in leaf litter for use by producers.

Decomposition: Its Variables

Decomposition is also influenced by a number of abiotic and biotic variables. Among these variables are moisture, temperature, climatic exposure, and types of vegetation.

Let's look at these environmental conditions and biotic variables in a **wetland** ecosystem. In a wetland environment , we see a wide range of decomposition occurring. The rate of decomposition is affected by the specific ecosystem. In the wetland there are very wet areas where the vegetation is partially or completely submerged. Other areas in the wetland are completely exposed to atmospheric conditions and are always dry. In both situations the decomposition rate is low. In addition, there are areas in the wetland that are both wet and dry. In this area, we find an increase in the growth rate of microbes, enhancing the rate of decomposition. **Aerobic microbes** (using O_2) have a higher rate of decomposition under moist conditions in the wetland ecosystem.

The importance of decomposition cannot be minimized. As you have noticed, decomposition puts back, into the circle of life, the nutrients and minerals needed for the existence of life on Earth. Decomposers become part of the recycling of nutrients and minerals as they grow, reproduce, and die.

Decomposers make a major contribution to any ecosystem. They put nutrients back into the soil in a usable form. These nutrients were once absorbed by living organisms and used in their life cycle.

We often think of the ill-effects of fungi and bacteria. Bacteria can cause many diseases such as strep throat and staph infections. Fungi can cause Athlete's foot and eczema, as well as all that nasty mildew growing in your bathroom and that hairy mold growing on your bread. However, remember, if we did not have these decomposers, we would be scaling mountains of dead matter.

Figure 2.27: WETLANDS. Vegetation in a wetland may be completely or partially submerged; however, some producers may also be completely above the water line.

ABIOTIC VARIABLES IN AN ECOSYSTEM

In review, we've covered the first two components of an ecosystem: **producers** (autotrophs) and **consumers** (heterotrophs). Within these two components we took a closer look at photosynthesis and decomposition. We have noted the necessity of each of these components as they affect the ecosystem.

We have seen one group of organisms receive its necessary requirements from another group and vice versa. One group depends upon the next. Matter is built up or made by producers, used by consumers, and broken back down by decomposers, all in the same ecosystem. Consumers determine how fast nutrients are recycled in the ecosystem and are therefore called **rate regulators**.

Now let's continue with the third component: **abiotic variables**. This component consists of the soil, sediments, organic and inorganic matter, and "litter", (decayed insects, plants, etc). All the dead or inactive organic matter comes from producer and consumer remains. Such organic matter is critical to the internal cycling of nutrients in any ecosystem.

Figure 2.28A: LEAF LITTER. What rate of decomposition would you expect in this environment?

Example of an Abiotic Variable

The make-up of the soil, for example, would have an affect on the decomposers in that ecosystem. A farm field which has been plowed under would have an abundance of organic matter and moisture. This would provide nutrients for a whole host of decomposers. A sandy beach would have less organic matter and moisture and, therefore, a lower population of decomposers. We are always interested in the quality of soil when we are gardening. Providing the necessary amount of nutrients for our crops is important. We depend upon the decomposers in the soil to cycle these nutrients back to the producers.

Figure 2.28B: AN ABIOTIC VARIABLE. The makeup of the soil, for example, has an effect on the decomposers in any ecosystem. This beach on the Gulf of Mexico has more organic matter and moisture than a sandy beach would, and therefore does have a high population of decomposers.

SUMMARY

Now you have the whole picture. Producers, consumers, and abiotic variables all work together in an ecosystem. Each component is dependent upon the other.

The driving force of the ecosystem is the energy of the sun. The sun regulates the input of energy as well as the outflow of energy in an ecosystem. One group of organisms receives its necessary energy from the previous group. This consuming and decomposing process continues through the life of the ecosystem.

The basic processes in a functioning ecosystem are photosynthesis and decomposition. Photosynthesis changes CO_2 and H_2O into carbohydrates (glucose). Decomposition returns nutrients to the ecosystem, releases energy, and converts organic matter into inorganic matter.

Remember our lily pad ecosystem? Can you name the "players" that might be involved in that system? Who are the producers? Who are the consumers? What are some abiotic variables? That small lily pad ecosystem is teeming with activity, isn't it? There are thousands of ecosystems on planet Earth, each unique, yet functioning in much the same way; however, no ecosystem can exist independently from its environment.

Figure 29: LILY PAD ECOSYSTEM. List the producers, consumers, and decomposers in this picture. What are the abiotic factors present?

✔ THINGS TO KNOW AND DO:

Define:

*rate regulators ____

____

____

*heterotrophs and autotrophs ____

____

____

*decomposition vs photosynthesis ____

____

____

List five abiotic variables of an ecosystem.

a.____

b.____

c.____

d.____

e.____

Discuss consumers as rate regulators.

➜ MODIFIED TRUE-FALSE: correct underlined word if necessary.

_________ 1.　The driving force of any ecosystem is the <u>sun.</u> __________________

_________ 2.　Photosynthesis changes O_2 and H_2O into carbohydrates. ________________

_________ 3.　Both organic and inorganic matter are <u>biotic</u> variables. ________________

_________ 4.　<u>Organic</u> matter is crucial to the internal cycling of nutrients in any ecosystem.

_________ 5.　Decomposers make a <u>minor</u> contribution to any ecosystem. ________________

✍ FILL IN THE BLANKS:

1.　Matter is _______________ ________________ by producer, _______________

　　by consumers, and _______________ _______________ by decomposers.

2.　Fungi can cause ______________ _______________ and ____________ in humans.

3.　Decomposers put back _________________ into the soil in a usable form.

4.　The basic processes in a functioning ecosystem are _________________

　　and ________________.

✎ IN SEARCH OF...

What would happen if we had no decomposers on Earth? Explain in detail.

3

ENERGY

ELEMENTS

ECOSYSTEMS

EMERGENCE

III. **ENERGY AND ELEMENTS IN ECO-SYSTEMS**

A.   Introduction/Definitions.

B.   Trophic levels

   1.   food chain
   2.   food web
   3.   pyramids

C.   Biogeochemical Cycles

   1.   hydrologic cycle
   2.   oxygen cycle
   3.   carbon cycle
   4.   nitrogen cycle
   5.   phosphorous and calcium cycle

D.   Summary

1.   Each time you think of producer, consumer, and decomposer, use the triangle:

2.   To help with remembering the six biogeochemical cycles, use our "cycle," the "PONCCH-mobile."

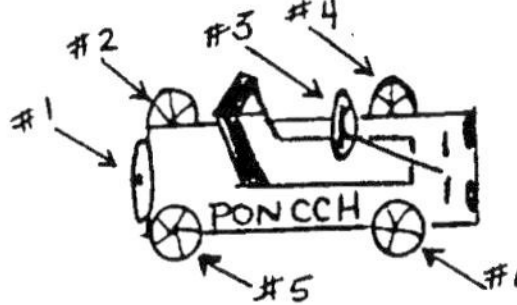

It has 6 "wheels"—each one is an element to remember—and they spell PONCCH—

P = phosphorus
O = oxygen
N = nitrogen
C = calcium
C = carbon
H = hydrogen

Things to Look for in This Chapter:

- Definitions of **all** highlighted words.
- How chemical energy moves from the sun to our dinner tables.
- How energy is lost on each trophic level.
- The difference between food chains and food web.
- How the natural elements in our biosphere cycle through ecosystems.

TROPHIC LEVELS/FOOD CHAINS PYRAMIDS

Ecosystems have many organisms living together within a specific area. Each organism has its own purpose. The producers, consumers, and decomposers in an ecosystem interact with one another in various relationships. Matter is built up, used, broken down, and recycled time and time again.

The sun's energy, through photosynthesis, is converted into chemical energy and heat. The sun begins this energy cycle, and the organisms in the ecosystems keep the cycle going.

It is worthwhile at this point to discuss food chains, food webs, and trophic levels and their importance to a well-functioning ecosystem. Feeding relationships among organisms fit in a specific spot or level in a **food chain.**

Figure 3.1: HUMANS AS CONSUMERS. Humans cannot produce their own chemical energy and various nutrients; they depend upon lettuce as well as other produce for food and energy.

Figure 3.2: A FOOD CHAIN. Food chains begin with producers such as grass and algae. Primary consumers such as fish and deer feed on the producers. Secondary consumers feed on primary consumers, and so on. Can you pick out secondary and tertiary consumers?

TROPHIC LEVELS

The term **food** refers to material that contains chemical energy for sustaining life. "Food" and "chemical energy" may be used interchangeably throughout this book. For all consumers, their source of chemical energy and nutrients is food that is needed for all their life processes: respiration, nutrition, digestion, excretion, assimilation, reproduction and circulation.

A **food chain** consists of a series of organisms feeding on one another. Food chains start with **producers** (green plants) who are eaten by primary consumers. These consumers are eaten by secondary consumers who, in turn, are eaten by tertiary consumers, and so on. A **food web** is a complex food chain. Most food chains and food webs are more complex than these definitions sound.

Figure 3.3: FOOD SOURCE. Grocery stores across the country provide a supply of produce, meats, cereals, and dairy products. Most consumers eat from all of these food groups.

Many different organisms in the food chain feed at more than one level. There is a complex flow of energy through this feeding process.

A hierarchy of energy transfers in an ecosystem is called **trophic levels.** Starting with the producers, energy is transferred to the next trophic level, the consumers. There are many types of consumers. Some consumers feed only on producers, while others feed on other consumers.

Those groups of organisms that directly feed on producers are called **herbivores.** Rabbits, various insects, and turtles are some examples of herbivores.

Carnivores are consumers that feed on other consumers and are therefore "meat eaters." Lions, owls, and snakes are some examples of carnivores.

Saprobes are decomposers which use dead matter for nutrition. Some examples are fungi, bacteria, and **detritus** eaters (organisms that feed on decaying material).

Humans, as well as other organisms seen in the food web, are able to eat at different trophic levels. Humans are called **omnivorous** because we are able to feed on both plant and animal matter.

Our financial situation influences the number of different levels from which we feed. Many people eat primarily at the lower trophic levels because of low economic status. The lower our economic level, the fewer trophic levels we use; however, at these lower levels, individuals receive more energy per unit of biomass than others receive at higher levels.

The energy that is lost during the passage to the next higher trophic level is an important factor in studying human population and human food supply. If we eat beef, much of the energy the cattle obtained from the producers (grass, alfalfa) is spent keeping warm and walking around. Not much energy is left for us. Greater use of energy by humans would occur if we ate from the producer level. We would be herbivores, occupying the second trophic level, rather than the third. Be sure to eat least 5 servings of fruits and/or "veggies" a day to obtain a greater amount of energy and nutrients.

Figure 3.4: FOOD WEB. A complex food chain is a food web. Look for overlapping food chains. Can you identify several of these in this food web?

Figure 3.5: HERBIVORES. A goat and addaxes are examples of herbivores found at Busch Gardens, Florida. They have flat teeth for grinding plant material. They are primary consumers.

Figure 3.6: CARNIVORES. Secondary and tertiary consumers such as snakes and tigers feed on other consumers.

Organisms at a given trophic level are at the same number of energy transfer steps away from the beginning energy source, the sun. (See Figure 3.7B) The energy transfer to the next trophic level is usually seen in a pyramid-shaped visual. The pyramid as a visual image in ecology was created by Charles Elton, an early ecologist. Today's ecologists are still using the pyramid to help students understand certain principles.

Look at Figure 3.7B, a pyramid showing the organisms at various trophic levels. All the organisms at the A level are producers. The level B organisms (primary consumers 1°) are one energy level away from the main energy source (sun).

Level C organisms (secondary consumers 2°) are two energy transfer levels away from the sun. The organisms at level D (tertiary consumers, 3°) are three energy transfer steps away, etc.

In Figure 3.7A, the sun is the initial energy source. The producers, using the sun's light, make up the first trophic level. **Producers** include algae, diatoms, seaweed, grasses, crop plants, trees, etc. The second trophic level organisms feed on the producers and are all

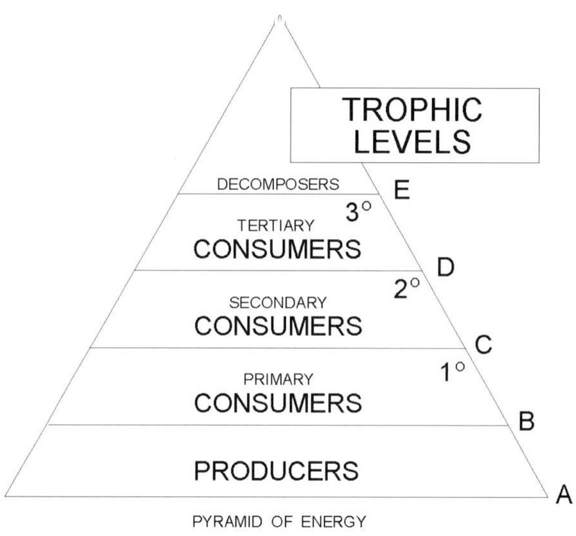

Figure 3.7B: PYRAMID OF ENERGY. Each step in this pyramid shows the amount of energy being transferred from one group of organisms to the next. Each step is termed a "trophic level."

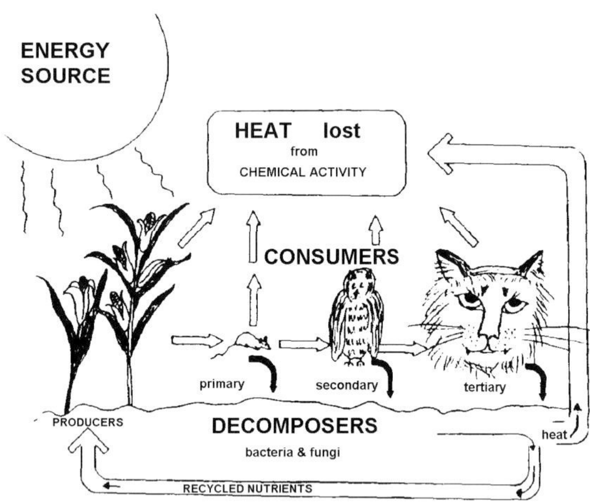

Figure 3.7A: ENERGY FLOW. The sun's energy is absorbed by the corn which is eaten by the mouse who, in turn, is eaten by the owl who is later eaten by the cougar. Then, his remains are broken down by decomposers; nutrients are released into soil for the corn's use.

Organisms found at each trophic level:

Producers
grasses, bushes, trees, algae, seaweed

Primary Consumers
insects, rabbits, deer, mice, cows, fish, manatees, giraffes

Secondary Consumers
owls, foxes, snakes, hawks, sharks, whales

Tertiary Consumers
lions, human, bears

Decomposers
fungi, bacteria

Figure 3.8: ORGANISMS AT DIFFERENT TROPHIC LEVELS. This chart lists organisms at each trophic level.

called **primary consumers**. This group includes insects, rabbits, birds, deer, giraffes, krill (tiny, shrimp-like organisms), aquatic insects, and various fish.

The next group of organisms in this hierarchy of trophic levels are the **secondary consumers** such as foxes, owls, whales, and spiders, which feed on primary consumers. **Tertiary consumers**, the next level, feed on the secondary consumers. Tertiary consumers include man, lions, tigers, and penguins. A lot of organisms, like man, can feed at different trophic levels.

We see a series of relationships developing with the transfer of energy at these trophic levels. Energy is lost at every trophic level in the form of heat, to some degree, as the energy is transferred from one feeding organism to the next. The decomposers then feed and cycle the once-fixed nutrients and minerals back into the soil for the producers' use.

The transfer of energy in ecosystems conforms to the **law of thermodynamics.** The first law states that energy is neither created or destroyed, but can be changed from one form to another. The second law states that with any energy conversion, some of the energy is lost to the system in the form of heat.

Figure 3.9: PRIMARY CONSUMERS. These giraffes feed on leaves (producers).

Figure 3.10: SECONDARY CONSUMERS. Foxes feed on primary consumers such as rabbits. (National Park Service).

Figure 3.11: TERTIARY CONSUMERS. Bears feed on producers as well as both primary and secondary consumers. (United States Forest Service).

Pyramids of energy represent the flow of energy from one trophic level to the next. Energy transference is not perfect; therefore, energy is lost to the environment all along the way. Due to the loss of energy, we see the characteristic shape of a pyramid being developed. From the base, each step away from the energy source gets smaller, thereby making the pyramid. It accurately reflects the laws of thermodynamics; this pyramid is right-side-up, with a large base at the bottom.

The following is an actual pyramid representation of energy flow at a fresh-water spring.

The base in Figure 3.12 represents the producers, eel grass and algae; the saprobes are fungi, bacteria, and detritus eaters. The herbivores are insects, snails, and turtles. The carnivores are insects and small fish. The top carnivores are large fish. We see in this figure, the small amount of energy at the top as compared to the producer level.

There are two more types of pyramids we will discuss. One is **pyramid of numbers**. It represents the number of individual organisms at each trophic level (see Figure 3.13). For example, in a square of grass, we would count the blades of grass, the various organisms living on the grass, insects, feeding on the grass, and so on. This data is then noted on the proper trophic level.

In Figure 3.13, the lowest tier represents the food production base for higher trophic levels of a bluegrass field.

This pyramid of numbers doesn't always adequately portray the trophic structure. It doesn't take into account that the sizes of organisms being counted in each trophic level can vary. That is to say that some organisms are very small while others are quite large; however, each is counted equally. A count in a redwood forest would yield a small number of large producers (trees) which support a large number of small herbivores and carnivores (insects); therefore, the pyramid of numbers would be "upside-down" (see Figure 3.14). Moreover, if the decomposers were also counted, they would outnumber other trophic groups by more than a billion.

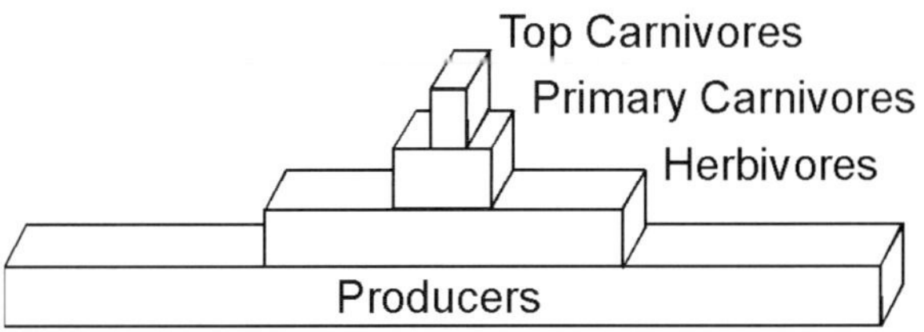

Figure 3.12: PYRAMID OF ENERGY. This is a pyramid of energy of a freshwater spring. Notice the abundance of producers compared to the remaining trophic levels.

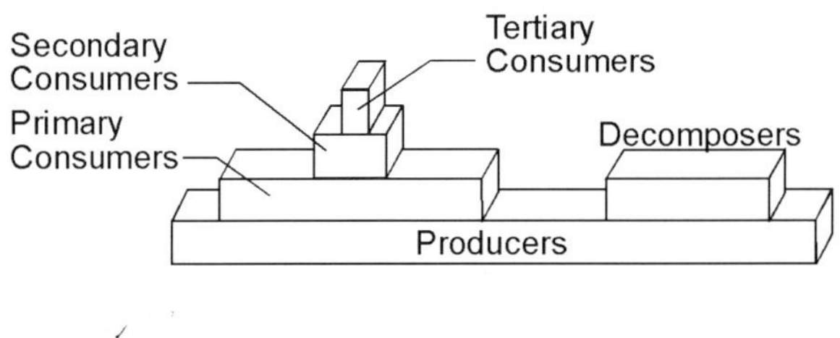

Figure 3. 13: PYRAMID OF NUMBERS FOR BLUE GRASS FIELD. Notice the large amount of decomposers in this field. Can you figure out what would be the tertiary consumers in a blue grass field?

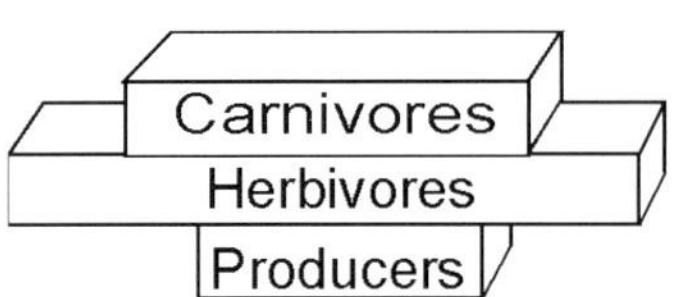

Figure 3. 14: PYRAMID OF NUMBERS IN A REDWOOD FOREST. Here individuals are counted. The large redwood trees, when counted, represent a small number in comparison to the abundant, but small, herbivores and consumers; therefore, the pyramid is inverted.

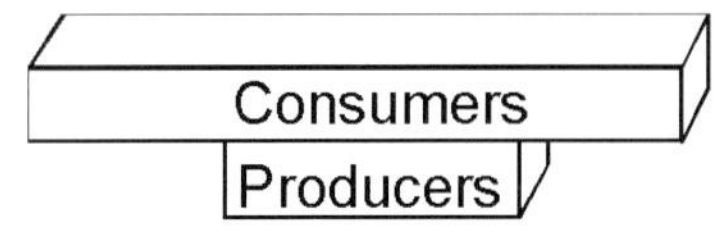

Figure 3. 15: PYRAMID OF BIOMASS IN AN AQUATIC ECOSYSTEM. Here the producers are more numerous in numbers but represent a smaller amount in biomass than the consumers.

The third pyramid we'll study is the **pyramid of biomass. Biomass** is the total dry weight of all living matter in an ecosystem. It represents the information collected on the amount of biomass in an ecosystem. This would be accomplished by harvesting a sample area, sorting out all the different organisms, and drying and weighing the matter collected. Again the data collected would be placed on the appropriate level of the pyramid.

For most ecosystems on land, such pyramids have a large base of primary production, with smaller and smaller trophic levels perched on top. In contrast, the producers of some aquatic ecosystems are tiny phytoplankton that grow and reproduce rapidly. Here, the pyramid of biomass can be upside-down, with the consumers biomass at any instant actually exceeding the producers' biomass, as seen in figure 3.16.

↘ DID YOU KNOW?

The amount of energy lost from one trophic level to the next is enormous. There is only about a 10% usable energy transfer between each trophic level. Much energy is lost as heat at each level.

↘ DID YOU KNOW?

Some plants eat animals. There are habitats where the chances for survival are very low because one or more factors may be too low or too high in quantity in a particular ecosystem. In a bog, for example, the element nitrogen is in short supply. Where there is a shortage, an organism will do what it can to survive. In North Carolina, Venus flytraps grow in bogs, a wet, spongy area. This plant has the ability to trap and digest insects. This added nitrogen source found in the insects' tissues is available to be used by the plant for its life processes.

Plants have several types of traps to capture their prey. Sticky substances on the hairs of sundew leaves are actually adhesive droplets to which an unsuspecting insect becomes stuck. Protein digesting enzymes are secreted by the plant. The plant absorbs the nourishment. Butterworts roll up their leaves slowly enclosing the insect within. The insect also was caught in the sticky substances found on the leaves.

Figure 3.16: VENUS FLYTRAP. Damselfly is walking on the inner portion of a venus flytrap leaf where the trigger bristles are located. The damselfly must touch two or three of these bristles to trigger the "capturing response."

Figure 3.17: VENUS FLYTRAP PLANT. This is an example of an insectivorous plant (flesh-eating). The toothpick is used to stimulate sensitive hairs on the surface of the leaf. The leaf has snapped shut tightly on the prey (toothpick). Digestion follows.

Bladderworts trap small aquatic insects who happen upon them. The insects are sucked into a bladder where a rush of water is released by the plant. One of the most famous of these flesh-eating plants is the Venus flytrap. This plant has hinged leaves. The edges of the leaves are toothed; the leaf blades contain sensitive hairs on the upper surface. An insect may land on a leaf of this plant and touch two or more of these hairs. The leaf blade then shuts in a split second. Its victim (i.e. ant) is held tightly within the folded leaf. Digestion may take ten to thirty-five days.

✔ THINGS TO KNOW AND DO:

Define:

* trophic levels ____

____

____

* biomass ____

____

____

* food web ____

____

____

* omnivorous ____

____

____

* food chain ____

____

____

* herbivores ____

____

____

* carnivores ____

____

____

* consumers

a.   primary consumers (1%)____

b.   secondary consumers (2%)____

c.   tertiary consumers (3%) ____

Give two examples of each of the following:

* primary consumers   _______________________   _______________________

* secondary consumers   _______________________   _______________________

* tertiary consumers   _______________________   _______________________

Create an original example of a terrestrial food web.

List 3 types of "pyramids" in ecology.

____

____

____

➡ MODIFIED TRUE - FALSE: correct underlined word if necessary.

________ 1.   The <u>producer</u> is the initial energy source. ____________________

________ 2.   A tertiary consumer feeds on <u>heterotrophs.</u> ____________________

________ 3.   A food web is a <u>simple</u> feeding process. ____________________

________ 4.   Energy is <u>lost</u> at every trophic level in the form of heat. ____________________

________ 5.   Primary consumers feed on <u>tertiary</u> consumers. ____________________

✍ FILL IN THE BLANKS:

1. _____________________, ____________________, and ____________________ interact with one another in an ecosystem.

2. ___________________is transferred at each trophic level.

3. ___________________consumers are at the top of the food chain.

4. A _________________ ___________________ is a complex food chain.

5. __________________ _________________ first used pyramids as visual examples in ecology.

✎ IN SEARCH OF...

Create an aquatic food web using all of the following:

whales, penguins, seals, Krill, lobsters, fish, man, squid, plankton, algae, crustaceans.
Your search may take you to other reference books.

BIOGEOCHEMICAL CYCLES

The energy and nutrients (elements) cycling through each of the trophic levels are absorbed, taken in, used, and released time and time again. We must consider how elements enter into, and exit out of, ecosystems. The elements required for life tend to move in cycles. The elements in these cycles are transferred from the environment to various organisms and then back to the environment. Our Earth serves as a storehouse of elements which are needed for the existence of all organisms.

For the remaining portion of this chapter, we will present and discuss the major cycles that influence an organism's survival. These cycles are termed **biogeochemical cycles.** In these cycles, nutrients essential for life are transferred from the environment to organisms and then back to the environment.

These elements are taken in by organisms as molecules and/or ions. Some elements remain unchanged; others become incorporated into new molecules, and still others are used for chemical energy. In the end, when waste or death occurs to the organisms, the elements reappear in the environment to be cycled once again. Decomposers play an important role in recycling these elements.

The following is a list of some of the biogeochemical cycles that will be presented in this chapter:

- hydrologic cycle
- oxygen cycle
- carbon cycle
- nitrogen cycle
- phosphorus and calcium cycle

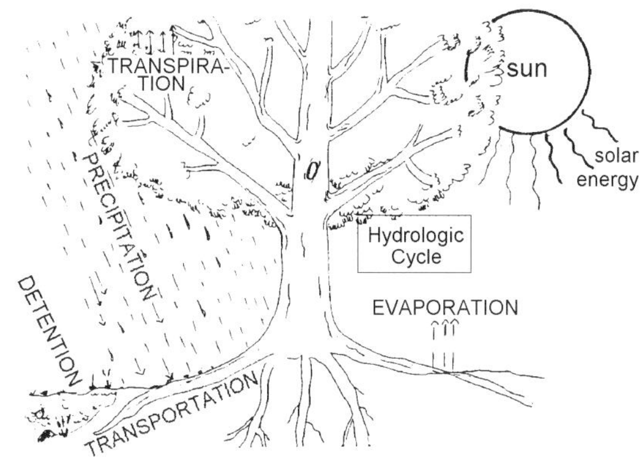

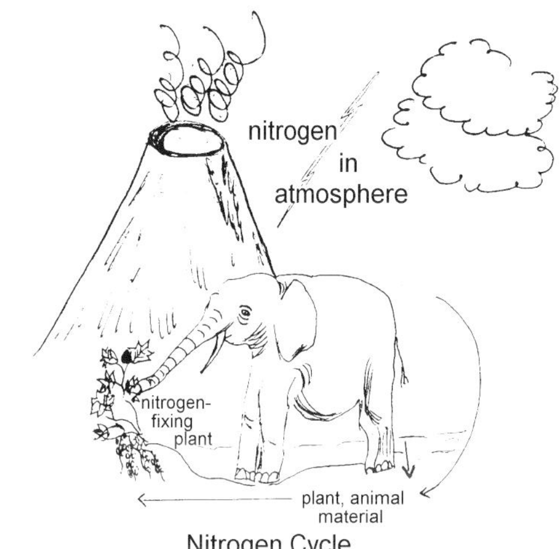

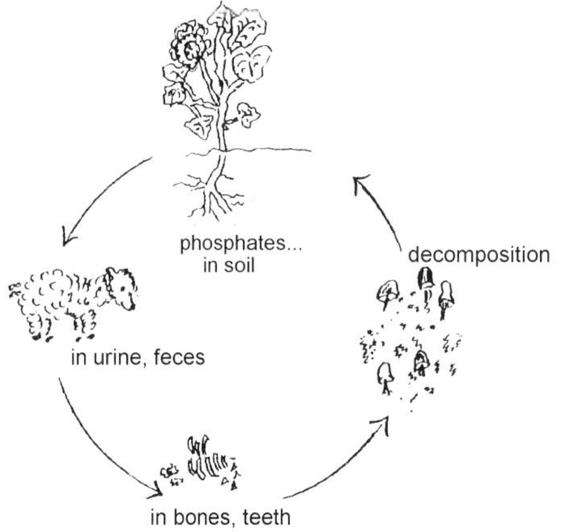

Phosphorus/Calcium Cycle

Figure 3.18: SOME BIOCHEMICAL CYCLES: HYDROLOGIC, NITROGEN, CALCIUM, AND PHOSPHATE. The elements in these cycles are transferred from the environment to various organisms and back to the environment.

Figure 3.19: HYDRO-LOGIC CYCLE. On a rocky beach, evaporation of water into the atmosphere takes place. Eventually, under the proper conditions, condensation of water vapor occurs and precipitation is later seen.

Hydrologic Cycle

The movement of water throughout the **biosphere**, which is the total region where life can exist on the earth, is called the **hydrologic cycle**. Life as we know it would not, and could not, exist without water. The biological processes that support life require the water molecule.

Water molecules do not usually stay in one place forever. They may be in a lake, river, or ocean, then evaporate into the atmosphere. They may fall back to earth as **precipitation**: rain, hail, snow, or sleet. Water may also become incorporated into plant and animal tissue.

Driven by solar energy, water moves slowly throughout the atmosphere, the upper layers of land, and the oceans, and then cycles back again. The cold and warm ocean current, wind, precipitation, and clouds are all part of the hydrologic cycle. Usually water in the atmosphere does not remain there for more than ten days; on land, the water molecule may remain for ten to 120 days.

The water molecules may travel from place to place by **evaporation, precipitation, transportation, transpiration,** and/or **detention.**

It is in **evaporation**, the release of water as vapor to the atmosphere, that the cycle begins. The major amount of evaporation occurs over the oceans since about 3/4 of the Earth's surface is covered by water. The water is returned to Earth as **precipitation**. Water may be stored in the land, in rivers, lakes, streams, ponds, and oceans.

Detention is the temporary storage of water on land or in oceans. The water may move on the surface of the land as **runoff**. This is called **transportation**. Water that has seeped into the soil and has been absorbed by the roots of plants enters the atmosphere through the process of transpiration. In **transpiration** water

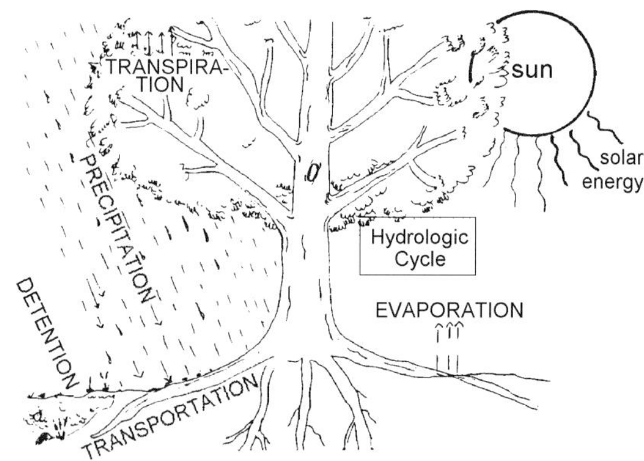

Figure 3.20: HYDROLOGIC CYCLE. Precipitation in any form is absorbed by soil and plants, while some results in runoff. Water is transported into and through the plant. Solar energy and photosynthesis cause evaporation and transpiration of the water, putting it back into the atmosphere.

vapor is released through tiny openings (stomata) on the underside of leaves.

In the 1950's water movements were studied in the White Mountains of New Hampshire. There, a small stream runs down the middle of a valley that has steep slopes and shallow soil. Several smaller streams are present, and they also run down the slopes of this valley.

We call this region a **watershed** because the rain falling on the land is funneled into a single stream. Watersheds are varied in size from a few acres to many hundreds of miles. The Mississippi River is an example of a watershed that flows for hundreds of miles.

The water that flows into a watershed comes mainly from snow and rain. Most of the water then filters into the soil. Some water may run off the land's surface into other bodies of water. The water in the soil may be absorbed by plant roots. Then the water is carried by the plant's vascular system (similar to our vein's function) to the canopy (tree tops). It is at this point, through transpiration, that water leaves the plant and enters the air once again.

This cycle continues; water from evaporation enters the atmosphere, falls to Earth as precipitation and is collected in watershed, absorbed by plants, transported from one area to another, then re-enters the atmosphere by evaporation. This is the hydrologic cycle. (See Figure 3.20).

Oxygen Cycle

Oxygen is an element which is also necessary for living things. In our atmosphere oxygen makes up about 21% of the total gas. It may exist as molecular oxygen and as ozone. Oxygen also exists in chemical combination in carbon dioxide and water vapor and is incorporated within living tissues.

We have discussed the oxygen cycle in the process of photosynthesis and in respiration in earlier chapters. Oxygen atoms are also part of water molecules and therefore part of the hydrologic cycle. About 90% of the oxygen in the Earth's crust is bound in calcium carbonate ($CaCO_3$) and is not available. In weathering of rocks, oxygen is removed from the atmosphere (oxidation) and is stored in the form of salts and oxides.

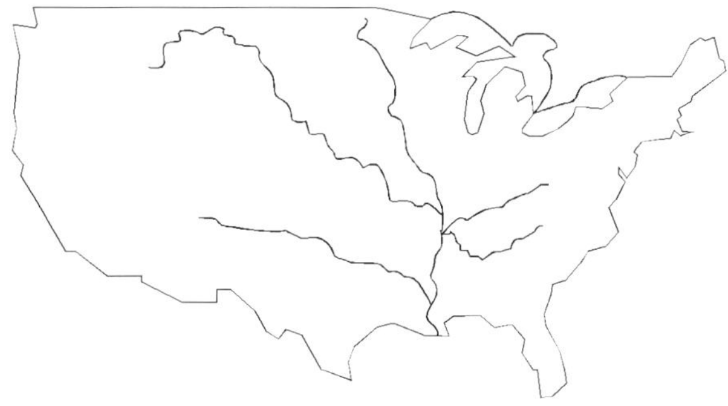

Figure 3.21: MISSISSIPPI RIVER WATERSHED. The Mississippi River is an example of a watershed that flows hundreds of miles. Many rivers and streams act as feeders for this watershed.

Ozone (O_3) at high altitudes is formed from molecular oxygen and atomic oxygen. We have extensively discussed the importance of ozone as it protects us from harmful ultraviolet radiation. The highly reactive ozone may combine with **fluorocarbons**, such as aerosol propellants, refrigerants, lubricants, solvents in making plastic, automobile exhaust, and other pollutants found in the atmosphere and may not be available in sufficient quantities to trap these harmful rays.

As discussed, the oxygen cycle incorporates both living and non-living substances such as calcium carbonate and ozone. The importance of oxygen cannot be overstated since our lives depend upon this element.

Figure 3.22: FLUOROCARBONS. Fluorocarbons such as those above have been found to be responsible for breaking down the ozone level and are now prohibited from use.

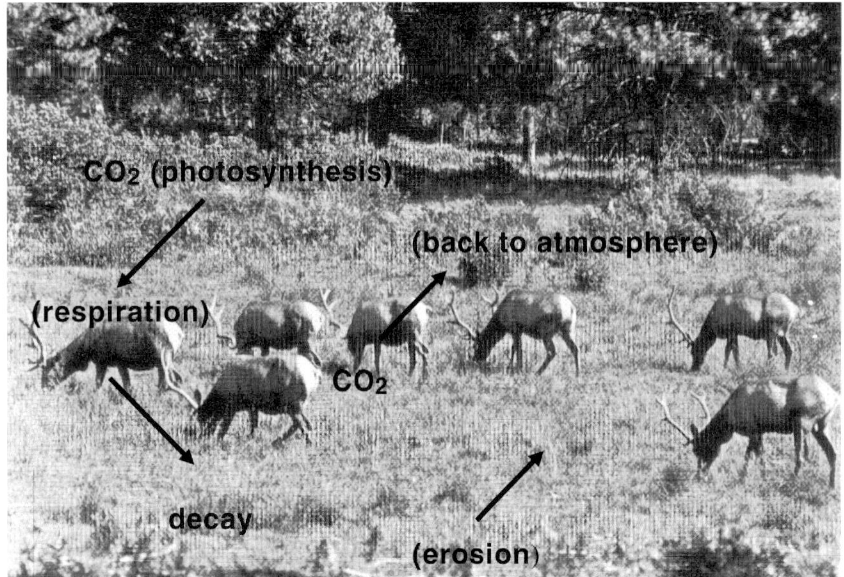

Figure 3.23: CARBON EXCHANGE. Carbon dioxide given off by the grazing elk is taken in by the producers (trees, grasses) for photosynthesis. (National Park Service).

Carbon Cycle

Carbon moves in a cycle from the atmosphere and oceans, through living things, and back. Carbon, through aerobic respiration, (through our exhaling, for example) burning of fossil fuels, and volcanic eruptions, enters into the atmosphere as carbon dioxide (CO_2), a gas. About 1/2 of the carbon is held within plant tissue and in oceans.

Through photosynthesis, CO_2 is taken in and incorporated into organic compounds that are used as a food source for consumers. The rate at which carbon remains in an ecosystem depends upon the type of ecosystem one considers. In tropical forests carbon uptake and decomposition is rapid, so the carbon is cycled rather quickly. The carbon is not bound for long periods of time in decaying plant matter (litter on soil surface). In ecosystems such as marshes and bogs, organic material is not broken down rapidly; therefore, the carbon stays bound for longer periods of time.

Carbon can be bound as carbonate in the shells of many aquatic organisms. When these organisms die, their shells become buried in the ocean floor for many years. Carbon is also slowly converted to gas, petroleum, and coal deep within the Earth and later tapped for use as fossil fuels.

Do you recall the discussion of the greenhouse effect in earlier chapters? Carbon has a lot to do with this situation. The amount of carbon going into the atmosphere is greater than the amount that is returned to the oceans and to plant tissues. This excess is due to the increased use of fossil fuels and other human activity. This interrupted balance in the carbon cycle has caused global warming.

The more we know, the more we should become aware of what we're doing to our environment and its effects on our world. We

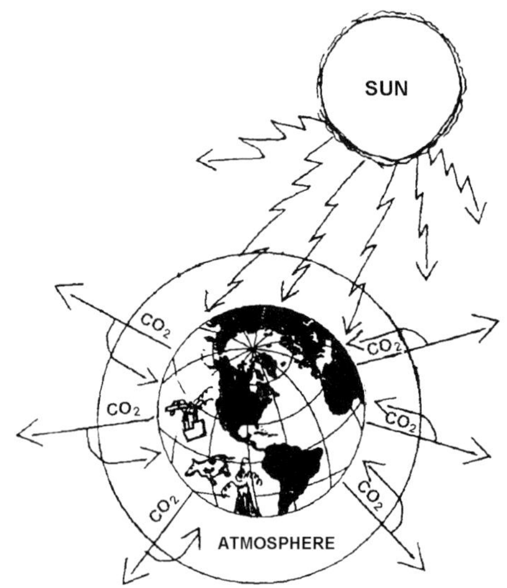

Greenhouse Effect

Figure 3.24: GREENHOUSE EFFECT. Excess carbon dioxide in our atmosphere retains heat for a longer period of time, thus causing an increase in temperature (global warming).

have put together the following list of possible events. These events may occur if global warming is left unchecked or ignored:

- Sea levels would rise due to higher temperatures warming the ocean surface temperatures (water expands when heated.).

- Partial melting of glaciers could cause coastal flooding world-wide.

- Melting of glaciers would also cause land masses to submerge.

- Weather patterns would change, influencing a decline in crop yields.

- Insect breeding may rise due to warmer temperatures, adding to even greater crop losses.

The carbon cycle exerts a tremendous influence on our lives and our future. Being ever-watchful and conscious of our environment and its delicate balance helps us become good stewards of planet Earth.

Nitrogen Cycle

Nitrogen is of great importance to the growth of land plants. It is found in proteins, nucleic acids (RNA, DNA), atmosphere, and

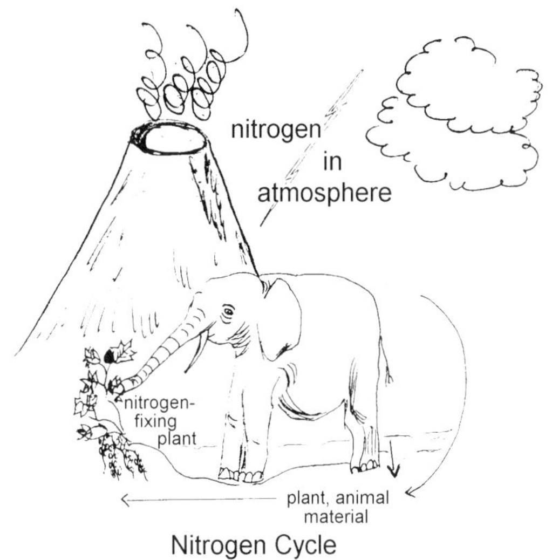

Figure 3.25: NITROGEN CYCLE. Through the action of volcanoes, lightning, decomposition, photochemical smog, and nitrogen-fixing bacteria, atmospheric nitrogen is converted to a usable form (available to producers to use).

oceans; however, not much of it is found in the Earth's crust. What little amounts of nitrogen found in our soil today comes from the action of **nitrogen-fixing organisms**.

We have mentioned in an earlier chapter that about 78% of the Earth's atmosphere is nitrogen (N_2). Only a few organisms in nature are able to use nitrogen in this form. Free nitrogen (N_2) does not readily combine with other substances. Some species of bacteria have the right type of **metabolism** (sum total of all life processes) for using atmospheric nitrogen (N_2).

One such nitrogen-fixing organism is a bacterium named **Azotobacter** which lives in the soil. Another nitrogen-fixing organism lives in a symbiotic relationship in the roots of plants; one example is **Rhizobium**. In aquatic ecosystems, **Anabaena** and **Nostic** are other organisms able to fix nitrogen.

Nitrogen-fixing bacteria are bacteria that produce nitrogen compounds by combining elements such as oxygen (O_2) with free nitrogen (N_2). These bacteria have the chemical ability to break down the triple bond that holds two nitrogen atoms together (N N). These

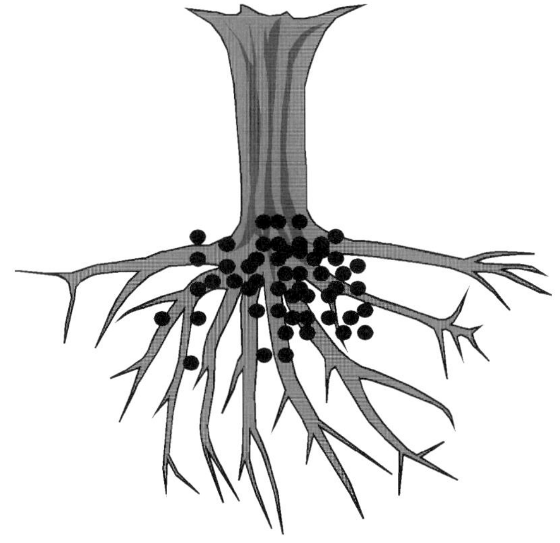

Figure 3.26: NODULES. Nitrogen-fixing bacteria living in the roots of legumes "fix" nitrogen for the plant's use.

nitrogen compounds are found in the proteins of plants and animals; later, these compounds are recycled back into the soil when the plants and animals die. The bacteria of decay come into play now and undergo a series of reactions that convert the nitrogen compounds into nitrates.

Besides nitrogen-fixing bacteria, there are other ways that atmospheric nitrogen (N_2) is converted into a usable form. Engines, volcanos, and lightning are a few of these ways.

The nitrogen cycle is comprised of six processes:

1. Nitrogen fixation
2. Assimilation - biosynthesis
3. Decomposition
4. Ammonification
5. Nitrification
6. Denitrification

In the **nitrogen-fixation** process there are certain nitrogen-fixing bacteria involved. These organisms live in the root **nodules** of

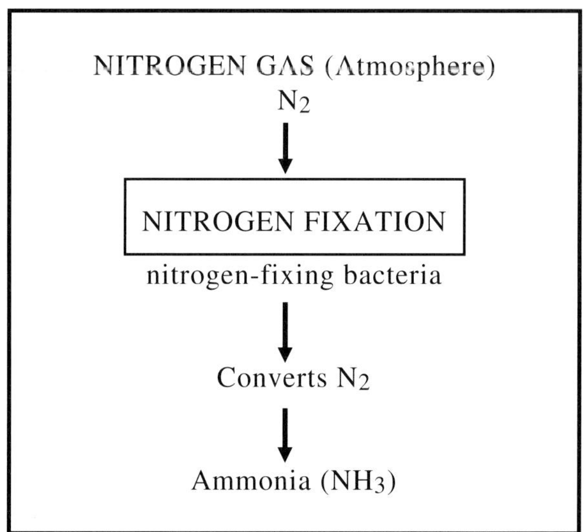

Figure 3.27: AMMONIFICATION. Decomposition of decaying tissue produces ammonia by the action of fungi and bacteria.

AMINO ACIDS:
Building Blocks of Protein

Those we synthesize:

1.	alanine	7.	serine
2.	proline	8.	tyrosine
3.	asparagine	9.	glutamic acid
4.	cysteine	10.	arginine
5.	glutamine	11.	histidine
6.	glycine		

Those we must get from our diet: "Essential Amino Acids"

1.	leucine	5.	phenylalanine
2.	isoleucine	6.	threonine
3.	lysine	7.	tryptophan
4.	methionine	8.	valine

Figure 3.28: AMINO ACIDS. The first group is a list of amino acids that our bodies are able to make. The second group of amino acids we must get from our diet since we are unable to make them.

plants called **legumes** (peanuts, peas, and beans). They can convert nitrogen in the air (N_2) to ammonia (NH_3). Once nitrogen has been converted to ammonia, it then can be used to produce amino acids, proteins, and nucleic acids.

The cycle continues as nitrogen is taken in by plants and incorporated into amino acids (biosynthesis). Chains of linked amino acids become proteins (assimilation). These proteins are nutrients necessary for growth and repair of an organism's tissue. **Nitrogen assimilation** is the process by which nitrogen is incorporated into organic molecules of living organisms (plants and animals).

The **decomposition** of decaying tissues and wastes produces ammonia. Bacteria and fungi are the agents of decay, producing the ammonia. Ammonia (NH_3) produced by decomposition is called **ammonification.**

Nitrogen may also be converted to nitrate ions (NO_3). This process, **nitrification**, also occurs during decomposition. For example, there are soil bacteria that convert nitrogen found in the proteins of manure to ammonia. Then other soil bacteria convert ammonia to nitrite (NO_2) and then to nitrate (NO_3).

Once the above reactions have happened, nitrogen is returned to the atmosphere through **denitrification** and leaching into water found in soil. This process is achieved by numerous bacteria as they reduce nitrate or nitrite to N_2 and a small amount of nitrous oxide (N_2O).

The continual production of ammonia by nitrogen-fixing bacteria in the nitrogen cycle leads us to assume that there is an abundance of nitrogen in the soil; however, soil nitrogen is scarce. The importance of this cycle, therefore, cannot be minimized or overlooked.

PHOSPHORUS AND CALCIUM CYCLE

These cycles involve nutrients in water instead of the atmosphere. Usable phosphates are found in soil water. Consumers receive phosphates through drinking H_2O and by eating from the lower trophic levels. Through decomposition, some phosphates become available. Other phosphates stay locked in animal remains, such as shell, bone and teeth. Erosion of these remains releases phosphate to the environment.

Phosphates are lost through leaching and run-off. Phosphorus enters the roots of plants as phosphate ions. Phosphate is used for the production of molecules such as ATP, a high energy molecule, nucleic acids, and coenzymes of photosynthesis and respiration.

Calcium is necessary for the proper functioning of cell membrane and many enzymatic reactions. Calcium enters the plant through its roots.

Calcium is bound in skeletons and shells. Leaching of calcium into the soil is a very slow process. A deficiency in calcium can interfere with the transport process and can cause the death of the plant. Calcium must come directly from the roots of plants and cannot be shifted from other plant tissues.

Both phosphorus and calcium lie bound in ocean floor sediments. Upwellings bring the sediments and dissolved ions to the surface where they are taken in by phytoplankton; then the cycle repeats itself.

These biogeochemical cycles can be put into three groups: hydrologic cycles, atmospheric cycles, and sedimentary cycles. In the **hydrologic cycle**, oxygen and hydrogen move in the form of H_2O molecules. In the **atmospheric cycle**, carbon and nitrogen, to a large extent, exist in a gaseous state in the atmosphere. Oxygen, at times exists in both the atmospheric and hydrologic cycles. In the **sedimentary cycle**, elements such as phosphorus and calcium remain within sediments found in the sea and in the land.

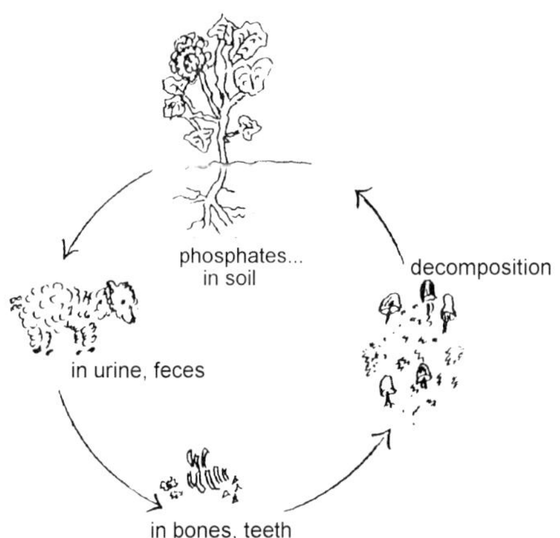

Phosphorus/Calcium Cycle

Figure 3.29: PHOSPHORUS AND CALCIUM CYCLE. Consumers receive phosphorus through drinking water and eating from the lower trophic levels. Calcium is obtained from plant tissue and dairy products.

SUMMARY

In conclusion, the elements required for life are transferred from the environment to organisms and then back again to the environment. These biogeochemical cycles occur in water, in the atmosphere, and in sediments. Some of the elements involved are oxygen, hydrogen, carbon, nitrogen, phosphorus and calcium.

Perhaps the next time you look at a tree, pass a farm, or water your favorite plant, you'll think about the hydrological cycle and the movement of water molecules. When you find a shell on the beach, you may think of the carbon found in that shell and how much time it takes to have that carbon released and where it goes from there. The knowledge you have gained in this chapter should give you a greater awareness and a deeper appreciation of planet Earth.

Figure 3.30: SHELLS ON A RIVERBED. Phosphorus and calcium remain bound in shell material for long periods of time and leach very slowly.

BIOGEOCHEMICAL CYCLES

Figure 3.31: BIOGEOCHEMICAL CYCLES. In this photo you will see all three types of biogeochemical cycles: hydrologic, atmospheric, and sedimentary.

✔ **THINGS TO KNOW AND DO:**

Define:

*biogeochemical ____

____

*hydrologic cycle ____

____

*detention ____

____

*transpiration ____

____

*stomata ____

____

*biosphere ____

____

*watershed ____

____

*nitrogen-fixing bacteria ____

____

*denitrification ____

____

Name at least 7 places you will find water molecules.

1. _________________________ 5. _________________________

2. _________________________ 6._________________________

3. _________________________ 7._________________________

4. _________________________

Explain how water molecules travel from place to place in the hydrological cycle. Fill in this cycle

1. ____

2. ____

3. ____

4. ____

5. ____

The nitrogen cycle is comprised of six events:

1. _________________ fixation

2. _________________ - biosynthesis

3. _________________

4. Ammon _________________

5. De _________________

Explain how the phosphorus and calcium cycles work.

____

____

____

> **→ MODIFIED TRUE - FALSE: correct underlined word if necessary.**

________ 1. The release of water as vapor by a plant into the atmosphere is called <u>evaporation</u>. ________

________ 2. <u>Detention</u> is the temporary storage of water on land or in oceans.

________ 3. <u>10%</u> of the oxygen in the Earth's crust is bound in calcium carbonate ($CaCO_3$).

________ 4. In the atmosphere oxygen makes up about <u>21%</u> of the total gas.

________ 5. Carbon enters the atmosphere as <u>carbon dioxide</u> (CO_2). ________

________ 6. <u>N_3</u> is free nitrogen. ________

________ 7. <u>50%</u> of the Earth's atmosphere is nitrogen. ________

_________ 8. Nitrogen-fixing bacteria are bacteria that produce nitrogen compounds by combining elements such as O_2 with <u>bound</u> nitrogen. _____________________

_________ 9. <u>Usable</u> phosphates are found in shell, bone, and teeth. _____________________

_________ 10. Phosphates are lost through leaching and <u>run-off</u>. _____________________

✍ FILL IN THE BLANKS:

1. Water vapor released through tiny openings on the underside of leaves is called

_____________________.

2. The Mississippi River is an example of a _____________________ that flows for hundreds of miles.

3. The movement of water throughout the biosphere is called the _____________________ cycle.

4. Molecular oxygen and atomic oxygen at high altitudes form _____________________.

5. Through _____________________, CO_2 is taken in and incorporated into organic compounds.

6. Carbon is not bound for long periods of time in _____________________

_____________________.

7. If global warming is left unchecked, sea levels would _____________________ due to _____________________ warming the ocean surface.

8. Also affected by global warming would be weather _____________________ and insect

_____________________.

9. Nitrogen-fixing bacteria that can convert N_2 to ammonia live in _____________________ of plants called _____________________.

10. _____________________ is the process of nitrogen being changed to ammonia.

✐ IN SEARCH OF...

Protein...where does it come from? Why do we need it?

____

____

____

____

____

BIOSPHERE

BIOMES

4

IV. THE BIOSPHERE AND BIOMES

A. The biosphere

    1. Man's influence in the bio-sphere

    2. Climate's influence
       a. solar radiation
       b. earth's rotation
       c. continent and ocean dis-tribution
       d. elevation of land masses

B. Major Biomes of the World

    1. Temperate forests
    2. Taiga
    3. Tundra
    4. Temperate grasslands
    5. Savanna
    6. Chaparral
    7. The Desert
    8. Tropical rain forest
    9. Aquatic

C. Summary

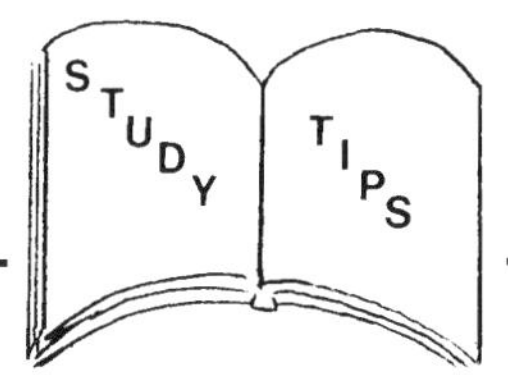

At the end of this unit you will need to know the nine biomes. Here's a memory tip. Use the following sentence to help:

(Ah, see the 5 T's)
A C D 5T'S

    A = aquatic biome
    C = chaparral biome
    D = desert biome
    T = temperate forest
    T = temperate grassland
    T = tropical forest
    T = taiga biome
    T = tundra biome
    'S = savanna biome

Things to Look for in This Chapter:

- Definitions of **all** highlighted words
- The differences between biosphere, hydrosphere, and atmosphere.
- How man and his technology influenced the biosphere in a positive as well as a negative way.
- How climate affects ecosystems.
- How climate affects biological rhythms.
- The names and characteristics of nine types of biomes.
- The types of flora and fauna that exists in each biome.

THE BIOSPHERE

Ecologists have come up with a single word to describe the living world in which we live, **biosphere.** The biosphere extends from the ocean depths to about 8 to 10 kilometers (5 to 6 miles) above sea level. The biosphere includes the terrestrial and subterrestrial environments as well as both fresh and marine waters.

The **hydrosphere** is the term used for the total of all water, frozen or liquid, on or near the Earth's surface. This would include oceans, seas, lakes, ponds, rivers, and streams, as well as ground water, polar ice caps, and airborne water (i.e. ocean spray, etc.)

The **atmosphere** is the area within the biosphere that consists of gases, airborne matter (dust) and water vapor surrounding the Earth.

The biosphere consists of all biotic and abiotic factors. Within the biosphere are numerous relationships of all types. Relationships may exist between two or more biotic factors, between abiotic and biotic factors, and even between the sun and the flow of its energy with both abiotic and biotic factors.

We as humans have relationships with both biotic and abiotic factors. One such relationship may be between corn plants and a farmer (producer-consumer relationships.) This relationship seems very direct as far as trophic levels are concerned, but let's look deeper into it.

Insects, fungi, nutrients, and weather are all factors which influence this relationship. Insects and fungi may attack the plant, lowering the food production. A nutrient deficiency in the nutrients in the soil may reduce the growth of the plant. Too much or too little rain would also hinder the plant's growth. All these factors cause a reduced harvest and an increase in the price of corn. This is just a small example of the many relationships that exist, and the ways they affect the biosphere.

Figure 4.1: BIOSPHERE. The biosphere includes all terrestrial, subterrestrial, aquatic, and atmospheric environments where life exists. (NASA).

Figure 4.2: HYDROSPHERE. Hydrosphere is the sum total of all water, gas, liquid, or solid, on or near, the Earth's surface.

Figure 4.3: ATMOSPHERE. The atmosphere is the area of the biosphere that consists of gas, airborne matter, and water vapor that surrounds the Earth.

Man's Influence in the Biosphere

Man, with his ability to control energy, has greatly affected the biosphere. His influence has consequently changed the web of life. As man gains more knowledge, develops new technologies, and finds new energy sources, his control of the biosphere becomes greater.

Our population continues to grow due to the increased control man has on his environment. This control allows a more favorable condition for the human species. Man now lives in areas of the biosphere that were previously not suited for life. Due to man's ability to control this environment, people now live on dry lands that have been irrigated, where deficient soils have been enriched, and where shelters have been built to protect from harsh climates.

Many problems and serious situations have also arisen from man's control and manipulation of his environment. Problems have developed from the growing populations, increasing pollution, and sometimes the uncaring, careless attitude of humans.

Take a moment or two to consider some situations which have been created by man. In the area in which you live, can you see some of man's negative and positive influences? Do you see relationships that have developed that have hindered or helped the biosphere? Can you trace the flow of energy from one trophic level to the next? How would you answer the question, "How has the biosphere where you live been affected by man?"

Figure 4.4: FARM TOOLS. Farming technology has improved over the years, allowing man to have more control over the results of his crop production.

Positive effect(s):

____

____

____

____

____

Negative effect(s):

____

____

____

____

____

Figure 4.5: MAN ADAPTS TO HIS ENVIRONMENT. Man has the ability to successfully live in almost all parts of the biosphere.

Climate's Influence

The biosphere includes ecosystems of all sizes, from the tiniest puddle to the largest forest. With few exceptions, all the ecosystems of the world are influenced by climate. **Climate** refers to the weather conditions that have been recorded over a period of time in an area. Temperature, humidity, wind velocity, degree of cloud cover, and rainfall are all part of the climate of an ecosystem.

There are many cities and states where climate has given these areas fame or at least notoriety: the windy city is Chicago, the sunshine state is Florida, and the states where "if you don't like the weather, just wait a minute; it will change," New England.

A climate is influenced by the following:

1. the amount of solar radiation,
2. the Earth's rotation (24 hours),
   the Earth's revolution (365 days),
3. the continent and ocean worldwide distribution, and
4. the elevation of land masses.

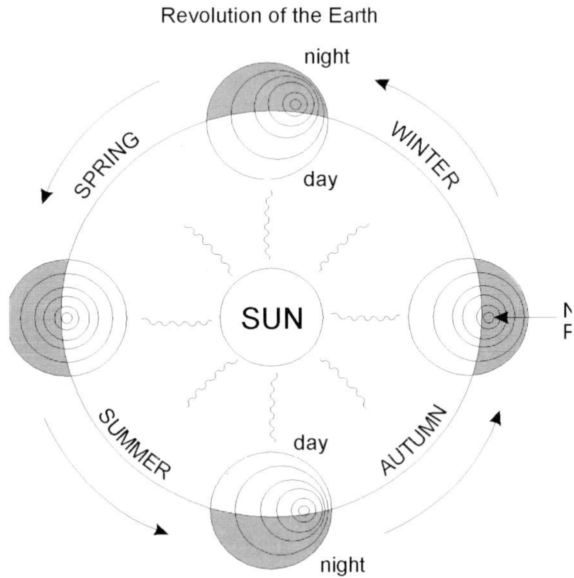

Figure 4.7: REVOLUTION OF THE EARTH. The above graphic shows the amount of sunlight reaching the Earth at different times of the years.

Yearly Precipitation (based on 1951-1990 data from Dept. of Commerce)

State	State where recorded	Annual Precipitation in inches
Alabama	Mobile	64.64
Alaska	Juneau	53.15
California	Los Angeles	14.85
Florida	Key West	39.42
Hawaii	Honolulu	23.47
Hawaii	Kauai	460.0
Kansas	Wichita	28.61
Nevada	Las Vegas	4.19
New Mexico	Roswell	9.70
Country	Station where recorded	Annual Precipitation in inches
Egypt	Cairo	1.1
China	Hong Kong	85.1
China	Singapore	95.0
Israel	Jerusalem	19.7
Peru	Lima	1.6
Philippines	Manila	82.0

Figure 4.6: YEARLY PRECIPITATION. The above chart shows yearly precipitation figures for several cities and countries of the world.

From the interactions of these factors, prevailing winds and oceans' currents are produced which influence global weather patterns. Weather patterns influence soils and sediments by erosion. These patterns also affect the composition and the distribution of soils, which, in turn, influence growth and location of specific producers.

Air currents influence the climate of an area because of the sun's radiation. Heat from the sun warms the atmosphere and drives the Earth's weather systems. At different latitudes the air masses absorb varying amounts of heat energy. This unequal heating of air causes the formation of air currents and global air circulation patterns.

These patterns cause regional variations in rainfall in the different areas of the world. For example, at the equator warmed air gives up moisture as rain when it cools at higher altitudes. This moisture supports lush forests. In contrast, air farther away from the equator descends at about 30º latitude and is drier. Deserts occur here due to the dry air.

Next, the air is warmed and picks up moisture and ascends at about 60º latitude. This traveling air mass moves toward the polar region. Here no moisture is given up as precipitation when the air mass descends. There is an absence of lush vegetation in this environment. In conclusion, the amount of rainfall in specific areas will influence the different types of ecosystems that develop.

The amount of sunlight falling on the Earth's surface is determined by the rotation and revolution of our planet. Our revolving Earth leads to the seasonal changes in the climate. Along with these seasonal changes many **biological rhythms** (recurring patterns) have developed. These rhythms are affected by temperature and precipitation fluctuations. The flowering of plants, breeding cycles, and yearly migrations of animals are examples of biological rhythms.

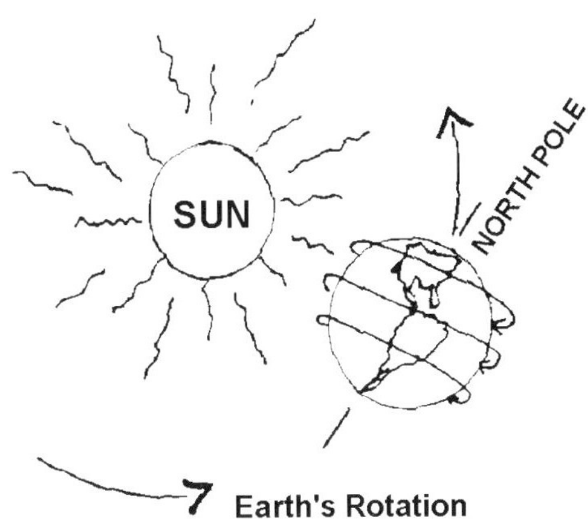

Figure 4.8: ROTATION OF THE EARTH. The graphic shows the rotation of our Earth as it revolves around the sun.

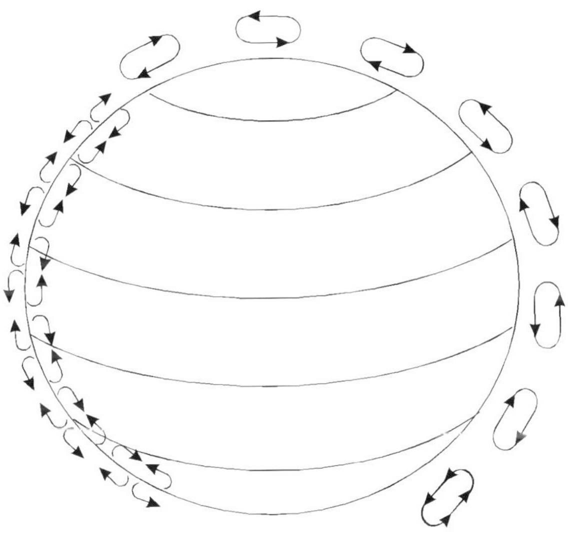

Figure 4.9: AIR CURRENTS CIRCLE THE EARTH. The graphic shows direction of air currents at various degrees of latitude.

One specific example of this type of rhythm is the length of darkness and its effect on a poinsettia plant. It will not flower unless it has a specific amount of uninterrupted darkness. So, when you plant your next Christmas poinsettia, you'll need to watch where you place it—**not** under your porch light that stays on all night.

Figure 4.10: BIOLOGICAL RHYTHMS. Poinsettias will not flower unless they have a specific amount of uninterrupted darkness. Another example of biological rhythm is yearly migration of different organisms.

The amount of rainfall and its effect on fruiting in a tropical area, and the drop or rise in temperature and its effect on migration of birds flying South for the winter are just two of the many other examples of biological rhythms.

The oceans' currents also play a role in the distribution of ecosystems throughout the biosphere. There are two immense circular water movements which occur in the Atlantic, Pacific, and Indian Oceans. There is an ocean current that moves warm, equatorial water to the north and south. Also, a current of water cooled at the poles moves towards the equator. These vast water movements shape the climate near the coast. Coastal cities may be covered by fog and have a milder climate as opposed to inland cities.

The relief (highs and lows) of the land (mountains and valleys) influences regional climate. As air rises, cools, and loses moisture in mountainous regions, vegetation differences are noticeable. At lower altitudes where semiarid conditions exist, the vegetation is shrublike. These shrubs are adapted to a low moisture environment.

As altitudes increase, more deciduous (oak, maple) and coniferous (pine, spruce) trees are present. In very cold environments at high elevations, evergreen trees, grasses, and sedges are found.

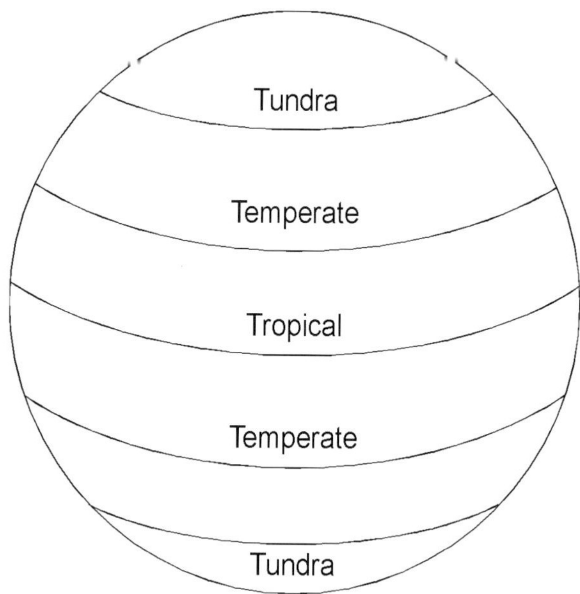

Figure 4.11: CLIMATIC ZONES. This simple graphic shows location of the different climatic zones of our world.

As the air descends the mountain, the temperature increases. This warmer air is able to hold more moisture than the cooler air at higher elevations. This warmer air actually draws moisture out of the plant life. The ecosystems here are semiarid. A reduction in rainfall on the leeward side of a mountain range is called **rain shadow. Leeward** refers to the

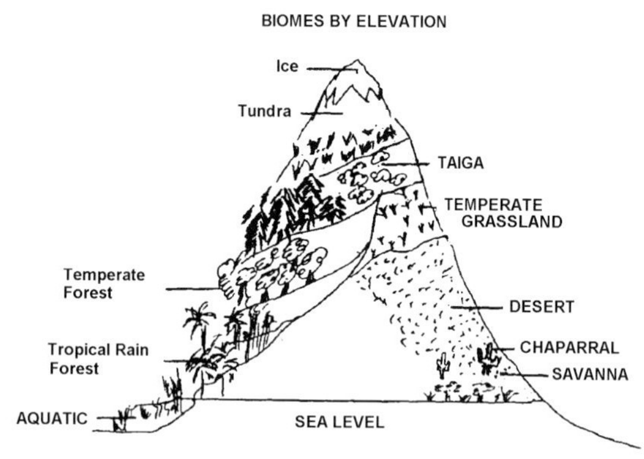

Figure 4.12: ALTITUDE AND VEGETATION. The graphic shows the location of Earth's biomes in relationship to elevation. Notice the type and location of organisms occupying the different levels.

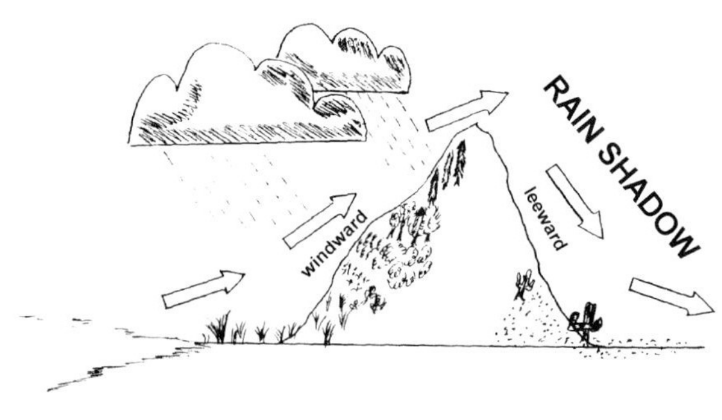

Figure 4.13: RAIN SHADOW. The reduced amount of moisture on the leeward side of a mountain is called rain shadow.

side of the mountain away from the wind; the **windward** side is the side that takes on the wind directly. The Himalaya Mountains of Asia and the Andes of South America experience rain shadow. In contrast, Mt. Waialeale on the island of Kauai in Hawaii experiences the increased moisture/rainfall because its windward side faces the sea where increased warm air combines with other geological factors causing the continuous rainfall.

Weather that occurs in an area is the climate. Climate is affected by the many factors that we have mentioned, interacting with one another. Variation in solar radiation, Earth's rotation and revolution, ocean currents, and the elevation of land masses affect the rainfall and temperature of ecosystems throughout the biosphere. With the existence of all these influences and their interactions, we have areas now in our world that are very different, different in climate, vegetation, animal life and land forms. Next we will present and discuss the major biomes of **our living world**, but first let's try some review questions.

Figure 4.14: RAINIEST SPOT ON EARTH. Mt. Waialeale on the island of Kauai in the Hawaiian Islands is the rainiest spot on Earth. At a height of 5,208 feet, the mountain has an average annual rainfall of 460 inches. (R. Naedele).

↘ DID YOU KNOW?

The rainiest place on Earth is in the United States. At the height of **5,208** feet, **Mt. Waialeale** on the island of Kauai in the Hawaiian Islands has an annual rainfall of **460** inches.

✔ THINGS TO KNOW AND DO:

Define:

*biosphere ____

____

____

*hydrosphere ____

____

____

*atmosphere ____

____

*climate ____

____

____

*biological rhythms ____

____

____

List the five elements found in a climate:

1. ____

2. ____

3. ____

4. ____

5. ____

The climate of an ecosystem is influenced by:

1. ______________________

2. ___________________ and ___________________

3. ___________________ and ocean ___________________

4. elevation of ______________________ _____________________

➡ MODIFIED TRUE-FALSE: correct underlined word if necessary.

________ 1. Due to <u>decreased</u> control of his environment, man's population continues to grow. __________________

________ 2. Weather patterns influence soils and sediments by <u>erosion</u>.

____

________ 3. Prevailing winds influence global <u>soil</u> <u>conditions.</u> ____________________

____

________ 4. <u>San Francisco</u> is known as the windy city. ____________________

________ 5. Global air circulation patterns are caused by <u>equal</u> heating of air currents.

____

________ 6. Temperature and precipitation fluctations affect <u>biological</u> <u>rhythms.</u>

____________________ ____________________

________ 7. A reduction of rainfall on the leeward side of a mountain is called

____________________ ____________________ .

________ 8.	<u>Deciduous</u> trees are pine or spruce. _____________________

________ 9.	Deserts tend to occur at <u>60° latitude.</u> _____________________

________ 10.	The amount of rainfall may affect _____________________ in tropical areas.

✍ FILL IN THE BLANKS:

1.	Circular water movements occur in the _____________________, _____________________, and _____________________ Oceans.

2.	_____________________ has control, to some extent, over his environment.

3.	At the equator, warmed air gives up water as rain when it _____________________ a high altitude.

4.	The _____________________ of plants, _____________________ cycles, and yearly migration are examples of _____________________ _____________________.

5.	_____________________ trees, like oaks and maples, grow as altitude increases.

✎ IN SEARCH OF ...

Explain how man's development of new technology has affected the biosphere. (Give at least 3 fully explained examples.)

____

____

____

____

____

____

____

____

____

MAJOR BIOMES OF THE WORLD

The biosphere is made up of several very distinct biomes. A biome is characterized by the flora and fauna within a specific area. The geographic area of the biome has been brought about by the distinctive climatic conditions which exist there.

Biomes are usually identified by the flora rather than the fauna because consumers depend upon the producers (flora). **Biomes** may be defined as groups of ecosystems that have similar types of vegetation. The major biomes are desert, tundra, temperate deciduous forest, tropical rain forest, aquatic, grassland, taiga, chaparral, and tropical savanna.

The area that you live in is one of the above biomes. Some people never leave their biome; others have chosen careers that take them from biome to biome. These world travelers spend time in biomes different from the one in which they live. They see the distinct differences and wonders of our diverse planet.

Figure 4.15: TEMPERATE FOREST. Climates here are mild, but may have some months that are very cold and some months that are very hot. (U.S. Soil Conservation Service).

However, as we move from biome to biome on foot, in a car, or on a train, we see no sharp differences or definite boundaries. Any area has gradual changes as we move across it. We hardly notice the differences. Plants appear in one form, then gradually become scarcer, with other plants popping into view.

Figure 4.16: TROPICAL RAIN FOREST. Here there is plenty of direct sunlight, warm temperatures, and daily rainfall. Much flora abound.

Figure 4.17: GRASSLAND BIOME. Grassland biomes are both flat and rolling and are usually found in the interior of continents. Both small and large animals may live here. (U.S. Forest Service).

When we discuss the many different biomes of the world, we have to omit many of the details due to the vast amount of material at hand. Each biome has its own particular set of characteristics; however, the change from biome to biome is gradual.

Between these distinct biomes are areas called **transition zones**. There, characteristics gradually end, changing old characteristics for new ones. Transition zones usually have groups of plants and animals specific to that area. No one biome, however, has a continuous, homogeneous vegetation cover.

When we limit and generalize a biome's characteristics due to the limited space available in the text, we omit and sometimes include only specific items. These generalizations and understatements may disturb acute observers of our world's biomes, but our purpose is to give you an awareness of the major characteristics of each biome of the world. Entire books have been written about each biome, but when you're finished with this "trip" through the major biomes, you will know a lot about God's green Earth. Sit back, relax, turn off the TV, and get ready for a trip around the world.

Temperate Forest

The first biome we come across is the **temperate forests**. These forests are located north and south of the tropics in areas called **temperate zones**. Climate here is mild, but may range from below freezing in some months to a sizzling 100° F in other months. Rainfall also varies in temperate areas from dry to moderately wet. Temperate forests exist where rainfall is between 75 to 150 cm (29–59 inches) per year.

Temperate forests are found in northern United States, Europe, Africa, Asia, and South America. (See Figure 4.18). **Evergreen** trees (conifers) and deciduous trees (shed leaves in fall) grow in temperate forests. **Conifers** have narrow, needle-like leaves that are not shed each year. Some evergreens are pine, spruce, redwood, and fir. **Deciduous** trees, on the other hand, have broad leaves that shed each year in the fall due to water loss. Some exam-

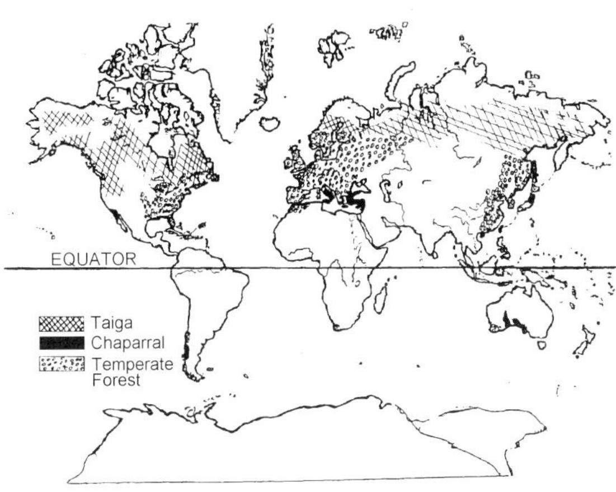

Figure 4.18: TEMPERATE FORESTS. Notice the location of several temperate forests worldwide.

TEMPERATE FOREST

Figure 4.19: FLORA AND FAUNA OF A TEMPERATE FOREST. The squirrel and oak tree are representative of the animals and vegetation in a temperate forest. Many large animals such as deer and bears live in these forests.

ples of broad-leaf trees are maple, oak, chestnut, hickory, birch, beech, and basswood. In the fall, before the leaves land on the forest floor, they turn bright orange, red, and yellow.

As producers, trees provide food for various forest creatures. These animals feed on bark, leaves, fruit, and seeds. Leaves and other tree parts which fall to the ground provide food for decomposers which turn decayed plant material into soil rich with organic matter.

In the United States and Canada, the mouse, grouse, snowshoe hares, and squirrels live in coniferous forests. In the ponds and streams you might see beavers and muskrats in their daily activities. Water fowl such as geese and ducks all make the forest their summer home.

In the deciduous forests of the United States, deer, bear, bobcats, foxes, and wild turkeys roam. Smaller animals like mice, voles, squirrels, chipmunks, raccoons, and opossum live on insects, mushrooms, nuts, and fruit. The larger animals feed on the smaller animals. Deer are usually found feeding on shrubs and seedlings at the forest's border. Beneath the groundlayer of leaves and other organic matter is the soil. It is a rich, gray-brown topsoil containing bacteria, protozoa,

Figure 4.20: DEER IN FOREST (photo on p. 6 of *Wildlife In America*). Many deer make their home in the temperate forest feeding on tender leaves and grasses. (National Park Service).

fungi, worms, and arthropods which all feed on organic matter.

Taiga

Our next stop is due north at the northern coniferous forests called **taiga**. It is there that long, severe winters and a constant cover of snow exists. Hemlock, pine, fir, and spruce are the evergreen trees found in this region. The thin, needle-like leaves of the conifers cover the ground along with fungi and dead twigs. Deciduous trees such as tamarack, willow, elder, birch, and poplar are found along the stream banks.

In this area, the larger animals are moose, black bears, and mule deer. The smaller animals are mice, snowshoe hares, shrews, lynx, and grouse. These smaller organisms feed on forest vegetation while the larger organisms feed upon mammals and/or leaves, buds, fruit, nuts, fish, campers' food, and tree bark.

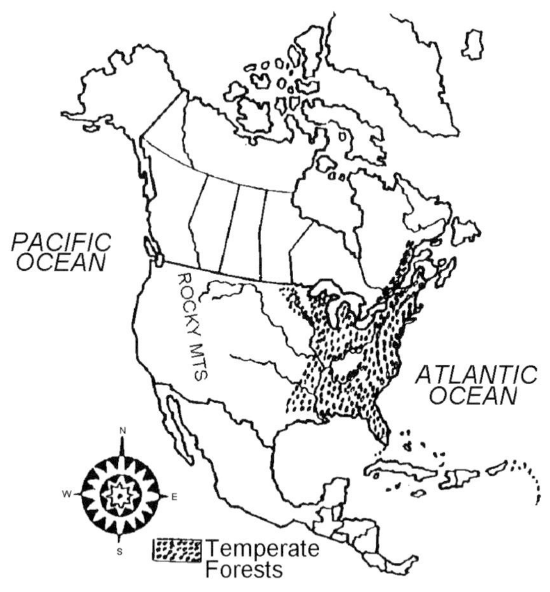

Figure 4.21: TEMPERATE FORESTS IN THE UNITED STATES. Most of the temperate forests are in the central and eastern parts of the United States.

TIAGA

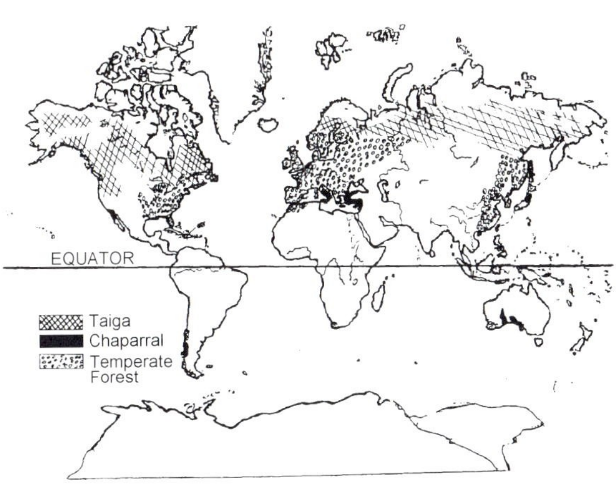

Figure 4.22: TAIGA BIOME IN THE WORLD. Notice the expanse of the taiga biomes across the North American, European, and Asian continents.

Figure 4.23: TAIGA BIOME FLORA AND FAUNA. The moose and poplar tree are representative of the animals and vegetation in a taiga biome. Many evergreens and small animals also live here.

Tundra

Our northern most stop is the **tundra**, where summer temperatures average between 0° to 45°C and winter temperatures range between -34° to -40°C. Summers here are short and mild, winters long and cold. Tundra is a form of grassland that occupies 1/10 of the Earth's surface. It forms a belt across Asia and North America. **Permafrost**, a layer of subsoil permanently frozen, is found in the tundra.

In the summer the ground may thaw a few centimeters, becoming wet and soggy. During the winter months this soil will refreeze. This process of freezing and thawing destroys deep roots; therefore, small, shallow roots may grow in the tundra. Due to the drying and abrasive winter wind and snow, only small, stunted plants survive.

The growing season is limited to about two months, from early spring until the first frost of fall. The vegetation is herbivorous, small shrubs, grasses, sedges, and rushes. A layer of moss and lichens lies beneath these plants. **Lichens** are a combination of algae and fungi

living on rocks and appear a crusty, grey-green color.

Flies and mosquitoes emerge during short Arctic summers. Migratory birds feast upon the large numbers of insects in the long daylight hours. To birds, who must eat constantly

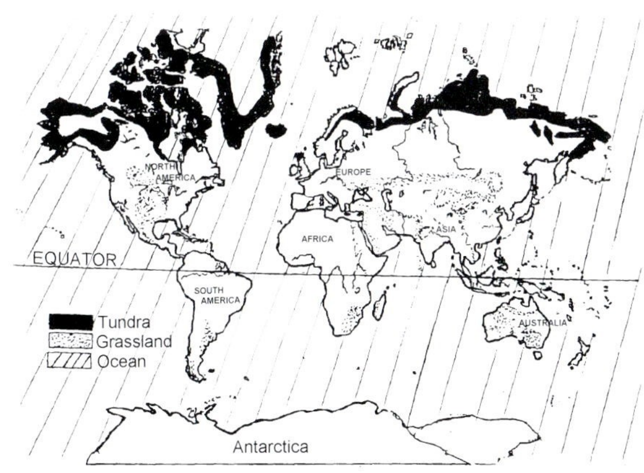

Figure 4.24: THE TUNDRA BIOMES OF THE WORLD. The northern most parts of the North American, European, and Asian continents make up the tundra.

due to their high metabolic rate, these summer months must seem like "hog heaven."

The tundra we just described is also referred to as the **"Arctic" tundra**. **"Alpine" tundra** also exists on high mountain peaks above 3,400 meters where trees do not grow. This height is above the **treeline**, the highest point where trees can grow. The weather conditions and the growing season are the same in both tundra areas.

Temperate Grassland

Now our trip stops in an area of transition between the temperate forests and the desert, the **grasslands**. Grasslands are usually found in the interiors of continents. Times of drought, hot-cold seasons, and grass fires are characteristics of grassland biomes. The terrain may be rolling or flat. Worldwide, the grasslands are found on the prairies and plains of North America, the **steppes** of Russia, the **veld** of South Africa, and the **pampas** of Argentina.

The flora of grasslands include grass (the dominant plant), legumes, and various annuals. Look at the grasslands in America. (Figure 4.26) The grasslands become more moist and have richer soil in the East.

The average rainfall for grasslands is between 25 to 75 cm (9-29 inches) per year. This

Figure 4.25: FLORA AND FAUNA OF A TUNDRA BIOME. The Arctic tern is one of the migratory birds that takes advantage of the long summer days in the tundra.

amount of rain is too high for a desert and too low for a forest.

The grassland worldwide is home to both large and small animals. Wild horses, bison, zebra, sheep, cattle, antelope, and gazelles graze on the grassland. These large grazers support the carnivores, lions, tigers, wolves, and humans. Grazing animals and fires maintain the grasslands by destroying tree seed-

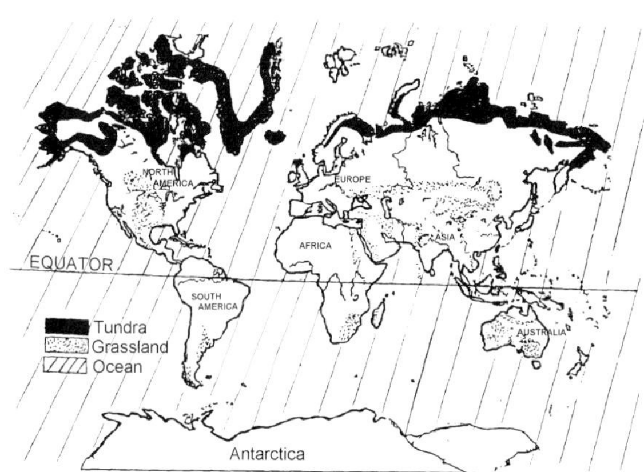

Figure 4.26: GRASSLANDS ON EARTH. Grasslands are found widespread, scattered throughout most continents.

GRASSLAND

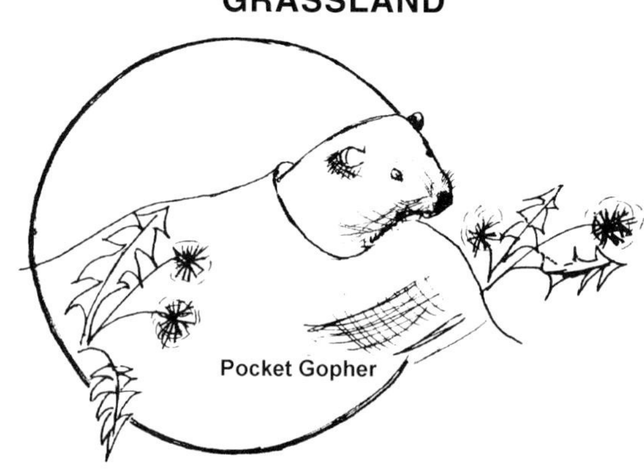

Figure 4.27: GRASSLAND FLORA AND FAUNA. Grasslands being transitional zones support many small and large animals and include a variety of grasses and annuals.

lings, preventing them from taking over the land. Large sheep and cattle ranches flourish there. Burrowing animals such as ground squirrels, prairie dogs, gophers, and jack rabbits inhabit the American prairie lands. Predators of these burrowing animals found on the prairie include hawks, owls, rattlesnakes, badgers, and foxes.

Savanna (Tropical Grassland)

Heading south, we reach the savannas of South America, Africa, and India. A **savanna** is a tropical grassland that is covered with grasses, shrubs, and a few scattered trees. Competition for water is critical here; grasses are well-suited to the fine, sandy soil. The dense, wide-spread root system of these grasses absorbs the maximum amount of water from the seasonal rains of the savannas. Vegetation flourishes during the rainy season, whereas, in the dry season, the aboveground portions of vegetation wither and die. In spite of this fact, the deep root system is able to survive the severest droughts.

There is a balance in the vegetation growth in a savanna. **Woody plants** (trees) die if there

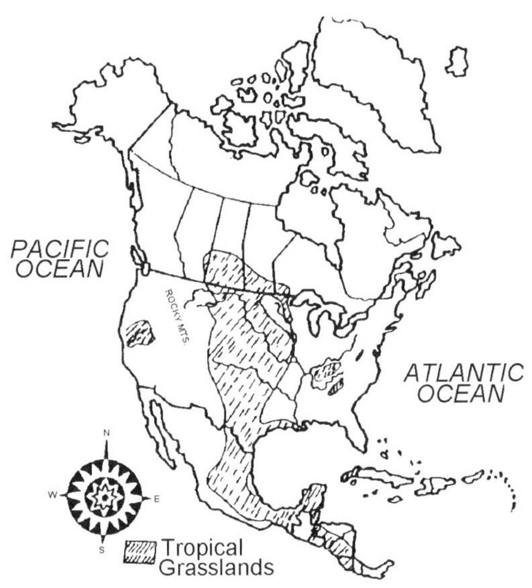

Figure 4.28: GRASSLANDS IN THE UNITED STATES, CANADA, MEXICO, AND CENTRAL AMERICA. Grasslands are transitional areas between deserts and forests and are usually found in the interior of continents.

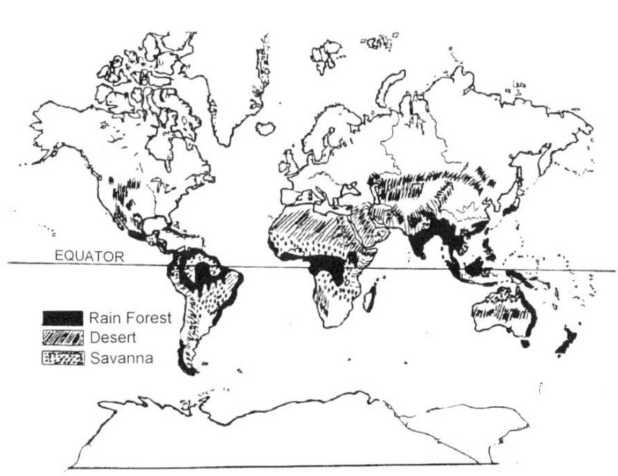

Figure 4.29: SAVANNA BIOMES IN THE WORLD. Savannas are tropical grasslands with scattered clumps of trees. Best known are those of Africa where they have a diverse group of herbivores.

Figure 4.30: FLORA AND FAUNA ON A SAVANNA. A black rhinoceros is one of the large herbivores that live in savannas. These animals are in imminent danger of extinction.

is a decrease in rainfall and grasses do well. If there is an increase in rainfall, trees increase in number until they shade the grasses. The shade of the trees causes the grasses to die. If a savanna is overgrazed, the moisture remains in the soil for a longer period of time. It is at this point that tree growth can flourish and the grassland is destroyed.

Big game animals and a wide variety of herbivores live in savannas. Gazelles, impala, iland, buffalo, giraffe, zebra, wildebeest, elephant, rhinoceros, and antelope also inhabit the savanna.

Predators such as lions and tigers live in these tropical grasslands. The predators mentioned, used the herbivores as their food source.

The rainfall here ranges from 100 to 150cm (39-59 inches) per year but is not spread evenly throughout the year. The dry season is long, compared to the rainy season. Grass fires are common during the dry times. If there is enough rain and tree growth is increased, the beginning of a tropical rain forest is established. The dry season with its low level of rainfall maintains a low tree growth; therefore, a grassland biome endures.

Figure 4.31: GRAZING ANIMALS OF A SAVANNA. A gazelle, another herbivore that inhabits the savanna, is a tasty meal for any of the roaming predators (lions, tigers).

Chaparrals

There are regions in Southern California, some Mediterranean communities, South Africa, and a portion of the Australian coast where the winters are mild and rainy, and the summers are long, hot, and dry. These areas are known as **"chaparral"** in California, **marquis** in the Mediterranean, and **matorral** in Chile. The climatic conditions of these areas

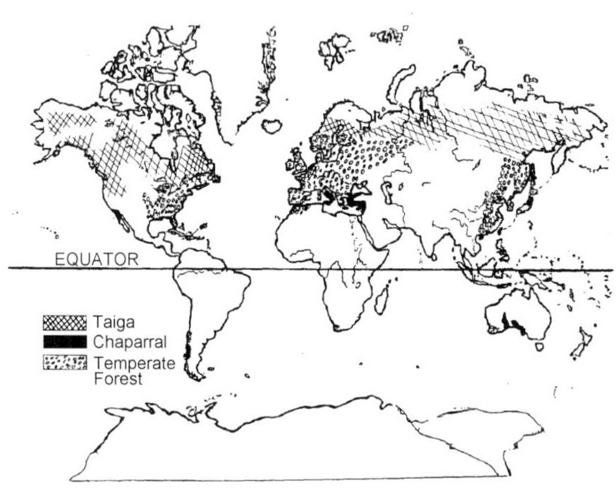

Figure 4.32: CHAPPARAL AREAS IN THE WORLD ARE FEW AND SCATTERED.

CHAPARRAL

Figure 4.33: FLORA AND FAUNA IN A CHAPARRAL BIOME. The mule deer, grasses, and shrubs are indicative of animals and plant life in a chapparal area.

are very close to one another; however, the plant life is different in species but do resemble each other in growth patterns and appearance.

The organisms living here are rabbits, lizards, wren, and towhees which are small, dull-colored animals matching the local vegetation. Mule deer in North America live in the chaparral during the spring season and move to cooler regions in the summer.

The Desert

Now, bring your sunscreen; we're heading for the desert. **Deserts** are areas of high temperature and low rainfall. They have less than 25cm (10 inches) of rain per year. The temperatures daily will fluctuate a great deal. In the daytime the temperature is around 38°C and then becomes quite cool at night due to the rapid loss of heat after the sun sets.

The largest desert is the Sahara which extends from the Atlantic coast of Africa to Saudi Arabia. The desert is about the size of the United States. (See Figure 4.35). It is expanding in size. The growth of the human population along the southern boundaries is the cause. The increased grazing of domesticated animals along the margins of the desert

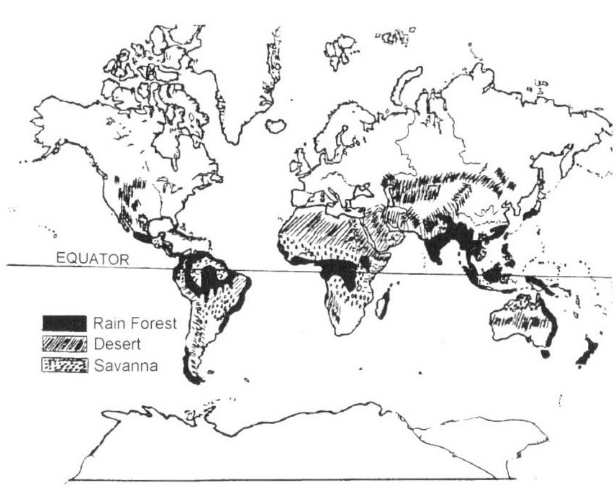

Figure 4.34: DESERTS OF THE WORLD. Deserts have very cool nights and extremely hot days—a great, daily fluctuation in temperatures.

is the single factor for this spreading. This overgrazing causes depletion of ground cover resulting in this increased desert area.

In North America, the desert is only about 5% of the total land area. The southwestern part of the United States is desert. This area is home to snakes, lizards, insects and birds. The

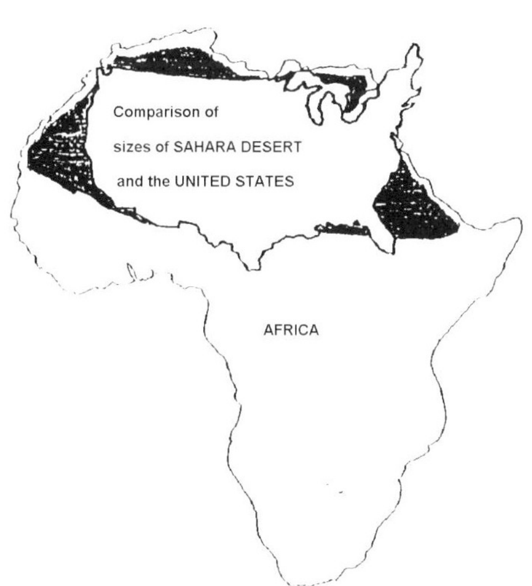

Figure 4.35: SIZE OF SAHARA DESERT. Does this give you a clear picture of the immenseness of the Sahara, the largest desert in the world?

Figure 4.36: FLORA IN DESERT BIOME. Cactus are succulents that are adapted for water storage because of the low annual rainfall in the desert.

desert animals are active during the cool night temperatures

After the spring rains, when enough moisture is present, the **annual** plants carpet the desert floor. Desert annuals are plants that live for one growing season, produce seeds, and then die. It appears as if there is a race during times of moisture, when a plant goes from seed to flower to seed again. **Perennials** are plants that have more than one growing season. These succulent desert plants are adapted for water storage. They store water in their roots and stems. Some of these desert plants may drop their leaves or have small leathery, water-conserving leaves. They also usually have extensive root systems that are able to trap large amounts of water when it's available. The root system goes down very deep into the soil, reaching the underground water.

Desert animals are also adapted to this climate by their waterproof outer coverings. Also, these animals conserve their water by having dry excretions. The mammal of the desert are few, small, and nocturnal. They obtain what water they can from the plants they eat.

Tropical Rainforest

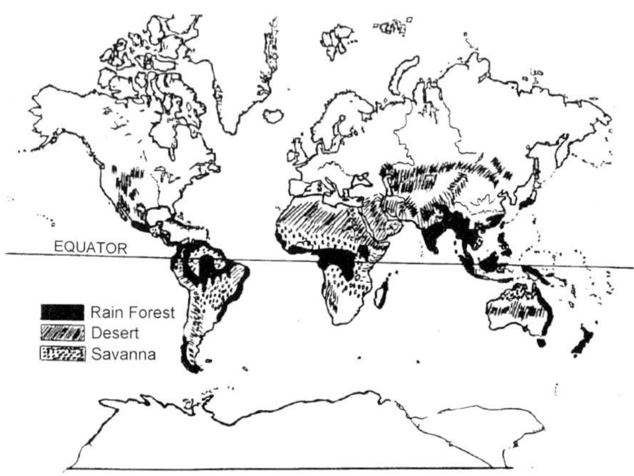

Figure 4.38: TROPICAL RAIN FORESTS. These forests are found in South America, Africa, and Southeast Asia. Most of New Zealand and part of Australia have rain forests.

Figure 4.37: FAUNA IN A DESERT BIOME. Most of the animals such as these lizards have a waterproof outer covering to limit water loss.

Rainfall is abundant in the **tropical rainforest** all year long, having between 200 cm - 400 cm (79 inches - 157 inches). There is plenty of direct sunlight, warm temperatures, and daily rainfall. From these climatic conditions, lush flora abound. It is asserted that there are more species of plants and animals in the tropical rainforest than in all other biomes of the world combined. There are many different species but low numbers of each specie.

The temperate deciduous forests have, in contrast, few species but high numbers of each specie. With the enormous varieties of species in a rain forest, it is difficult to identify the dominant plants.

Epiphytes, aerial plants, are quite common in the rainforest. Orchids, one aerial plant, lives on the limbs and trunks of trees. The epiphytes obtain water from the humid air and rainfall. These organisms need to conserve water and in some cases resemble the desert succulents. They have water-storing leaves and stems, spongy roots, or cup-shaped leaves that capture water. Other epiphytes include ferns, mosses, and bromeliads. Many vines are also seen in the rainforest. The woody vines, or **lianas**, are everywhere and have measured up to 240 meters (788 feet) in length.

Figure 4.39: FLORA IN A TROPICAL RAIN FOREST IN HAWAII. Orchids are epiphytes, one of the aerial plants living on limbs and trunks of trees. (Dr. RonnAnn Naedele).

The upper canopy exists of trees 50 to 60 meters tall. (about 164 feet - 197 feet) They have thin bark, wide, leathery, dark green leaves, and their root system is not very deep because the ground is always wet. The lower canopy is formed by trees of less heights. At ground level are the plants that are able to survive with low light intensity, like the African violet.

Figure 4.40: WOODY VINES IN A TROPICAL RAIN FOREST. Woody vines, or lianas, may grow over 700 feet in length.

Figure 4.41: BROMELIADS IN A TROPICAL RAIN FOREST. Another epiphyte, bromeliads, are found in the tropical rain forest and are adapted for collecting and storing water.

RAINFOREST

Figure 4.42: TYPICAL FLORA AND FAUNA IN A RAIN FOREST. An anteater looks for lunch in a tree covered with air plants.

Decomposition of matter is rapid and the competition for the nutrients is high. It has been noted that anything that falls to the ground is immediately carried off, consumed, or decomposed.

With this rapid uptake of nutrients, the soil in the rainforest is not very fertile. Leaching by heavy rain contributes to the low soil fertility. The soil where forests once were is poor today.

After a few year's use, increased upscale clearing and cultivating is destroying the rainforests at a tremendous rate. The predicted rate of the disappearance of the rain forests is as soon as the year 2000. As the rainforests are destroyed, the diverse variety of flora and fauna also disappear.

What we all can do to help keep this terrible disaster from happening is a hot topic and a concern to all of us. We must remember to be good "stewards", even if it is not in our own back yards.

The animal life in the tropical rainforest is diverse. Great numbers of species of colorful birds, insects, lizards, monkeys, and other

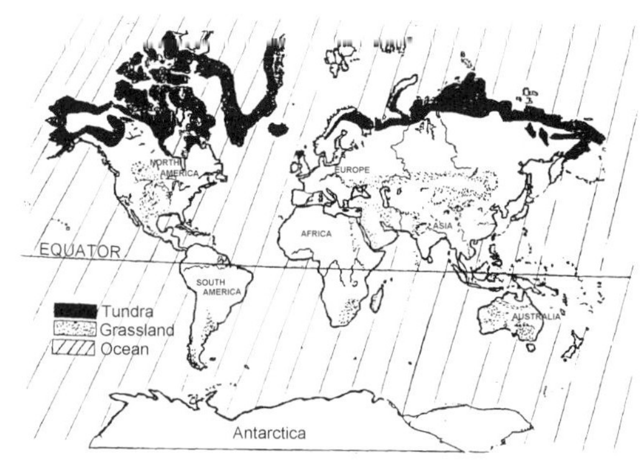

Figure 4.43: OCEANS ON EARTH. Seventy-five percent of our Earth is covered by water.

small mammals are found there. Most animals in the rainforest live in the trees.

Aquatic

Our journey continues as we move to the vast waters of our world which cover about three-quarters (3/4) of the Earth's surface. This **aquatic biome** contains more plants and animals than found on land. Most of the water

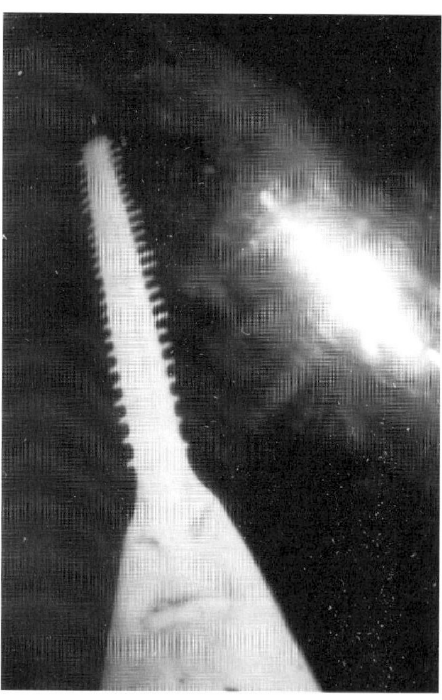

Figure 4.44: FAUNA IN OCEAN BIOME. An eerie fellow to meet up with in the depths of the ocean is the swordfish.

Figure 4.45: FLORA AND FAUNA IN AN OCEAN BIOME. There is a wide variety of aquatic marine life, both flora and fauna, in our oceans.

we encounter is salty and is called **ocean**. Other waters are fresh; they are the lakes, streams, ponds and rivers.

People are somewhat fascinated by the aquatic environment. The terrific underwater world is teeming with brightly colored fish, seaweed, and coral. Amazing and very strange creatures of the deep are intriguing and unique. It's hard to imagine life without those exciting and scary tales of mermaids and sea monsters.

Putting mermaids and monsters aside, we'll now explore the aquatic environment. First is a list of ocean characteristics:

1. Oceans hold a consistent supply of dissolved salts and nutrients.
2. Oceans are the most stable aquatic environment.
3. Oceans absorb and hold large quantities of solar radiation; this helps stabilize the temperature of the Earth.
4. Oceans are the most populated places on Earth.

There are about 3.45% dissolved minerals in the ocean. **Sodium chloride,** salt (NaCl) comprises just over 1/2 of the minerals found here. Ocean water also contains oxygen (O), potassium (K), chlorine (Cl), sulfur (S), magnesium (Mg), calcium (Ca), and sodium (Na).

The oceans are divided into areas where light can and cannot reach. The **littoral zone** is along the shores of the ocean. Light is plentiful here and supports the abundant **plankton**. Plankton are microscopic organisms (algae, diatoms) that are the basic link of the ocean's food chain.

The main factors affecting the life of aquatic organisms are:

1. the quantity of available oxygen and carbon dioxide,
2. the temperature
3. the presence of dissolved or suspended materials
4. intensity of light

Sunlight reaches down to the ocean floor in the littoral zone. Where light penetrates, producers are able to undergo photosynthesis (see Figure 4.46).

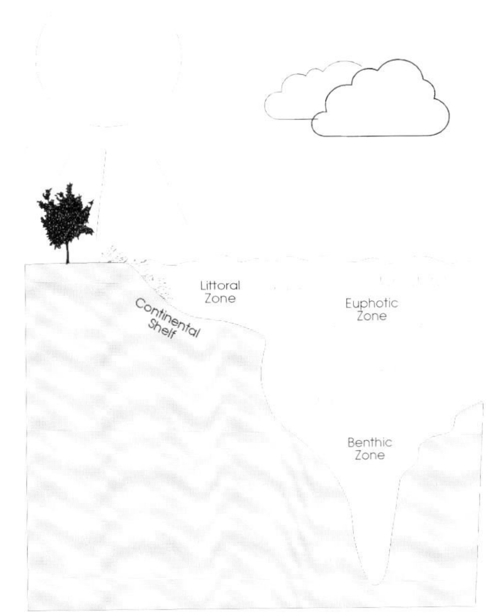

Figure 4.46: VARIOUS OCEAN ZONES AND ORGANISMS THAT LIVE THERE. The amount of light reaching different levels in the ocean dictates the life that can exist there. Some marine organisms can exist in different zones. (C. Sherman)

Plants in the sea are usually limited to the upper levels where light can penetrate, the **euphothic zone**. We see seaweed growing along the shoreline attached to the bottom's substrate (things on ocean floor) as well as floating along the top layers of ocean waters.

Microscopic organisms called **phytoplankton** are spread across the entire ocean surface. The growth of plankton is abundant where there are sufficient amounts of dissolved minerals and carbon dioxide. The areas of highest nutrients and carbon dioxide would be near the shoreline where runoff carries nutrients from the land. It is estimated that about 90% of the food making and oxygen releasing on the Earth happens in the water.

Another characteristic of oceans is their relatively stable environment. The sea is less variable than our terrestrial biomes. Sea water is fairly constant at pH8.2. Because of the high salt content, **salinity,** of the ocean water, it does not freeze at 0º (32ºF). The yearly water temperature ranges from about 27º C (81º F) to 1º C (34º F). The yearly temperature range of the ocean is minimal when compared to terrestrial environments, which may range from a sizzling 58ºC (136ºF) in the desert to a "chilly" -68ºC (-90ºF) in Siberia.

Figure 4.47: ALGAE (SEAWEED) WASHES UP ON SHORE. When looked at carefully, one will notice a wide variety of aquatic organisms inhabiting the seaweed.

DID YOU KNOW?

The Dead Sea is so salty that everyone "floats" on top of the water.

As far as population is concerned, animals live in all parts of the ocean, even in the deep Mariana Trench. (See Figure 4.48), a depth of 11,000 meters (36,000 feet). Each species has its own habitat in the ocean. An example of this is the habitat of flounder which live on the ocean floor of the North Atlantic coast. The Pacific Ocean off the coast of Washington is the habitat of salmon. The California sea lion lives off the California coast. Clams live in muddy beaches while sea urchins live in kelp beds (large seaweed).

The oceans present its inhabitants with certain factors that other organisms never meet. Differences in pressure are enormous from the surface waters to the ocean deeps. Great atmospheric pressure can limit life in the deep parts of the ocean. Some invertebrates are found living in the deepest parts of the **Benthic zone**, (deep ocean zone) where the pressure is 1,000 atmospheres (about 15,000 lbs. per square inch.)

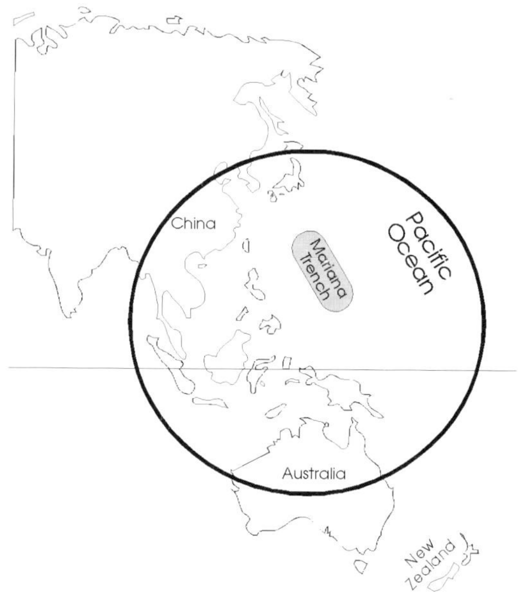

Figure 4.48: MARIANA TRENCH. The deepest trench of the ocean, where the pressure is severe and the organisms are very exotic and strange, is in the South Pacific Ocean.

This pressure is caused by the weight of ocean water. In order for organisms to survive, they must be adapted to the pressure. When organisms from the deeps are brought up to the surface, they do not survive because of the drastic pressure differences that exist between the ocean floor and ocean surface. Deep sea fishes are quite small and have large mouths. Most of these fish feed on organic matter which sinks down from upper levels.

Crustaceans, crabs, shrimp, lobster, and barnacles are as numerous in the ocean as insects are on the land. Fish are the most numerous vertebrates in the ocean. Herring, mackerel, tuna, and sardines are familiar commercial fish in the euphotic zone. In the benthic zone, fish appear very strange to our eyes. Some have huge eyes, long, thin tails, glowing spots, and wide mouths. In the deepest, darkest, coolest areas, fish, sponges, and echinoderms have been found.

The ocean coasts have populations that are under the water during high tides and exposed to the air during low tides. These populations of marine organisms have had to adapt in a different way from those interesting looking creatures of the deeps. They have special adaptations that allow them to survive both in water and on land between tides such as a clam burrowing into the sand during low tide.

Other aquatic biomes, ponds, streams, and rivers, support algae, frogs, salamanders, snails, and flatworms. Plankton are also found

Figure 4.49: FAUNA IN A FRESH WATER BIOME. Carp feeding in a pond in Central Florida display their various coloration and size in their species.

in fresh water ponds, streams, and lakes, providing food for larger organisms. Freshwater plankton include algae, protozoa, rotifers, and freshwater crustacia.

The temperature of lake water varies more than ocean water. Seasonal atmospheric temperatures cause lakes in the spring and fall to have an increased temperature range from the surface to the lake's bottom. The water over-

Figure 4.50: TIDAL MARSH. A raccoon is looking for a tasty marine morsel at low tide in a salt-water tidal marsh.

turns, bringing the oxygenated surface water to the bottom and the nutrient rich sediment water to the top. Plankton feed on this surface water and bottom dwellers use the oxygen in everyday survival.

In fresh water biomes organisms have adapted to pollution, to differences in light intensity and temperature fluctuations, to concentrations of dissolved gases and suspended particles, and to speeds of currents. However, unlike oceans, fresh water biomes are not stable. They vary in size from time to time and are subject to **succession**, organisms replacing organisms over a period of time. We will discuss succession in detail in a future chapter.

Aquatic biomes are diverse in type and in the organisms they contain. We are to be thankful for the oxygen, food, and recreation we receive from the waters of the Earth. Being ever mindful of what we can do to prevent pollution and destruction of our water environment is everyone's concern and responsibility.

SUMMARY

Man's influence on the biosphere has spread to every corner of every biome. He has left both positive and negative "footprints" in every environment. Besides man, climates have affected and shaped various biomes. The diversity of these biomes lets many organisms that are different from one another exist in similar environments. These differences are limitless. Experience all that you can by stopping, looking, smelling, and listening to the "living world" around you.

DID YOU KNOW?

The Sargasso Sea is covered with **sargassum,** a floating seaweed, which covers surface layers. This floating seaweed is a habitat for many marine organisms.

Figure 4.51: FRESH WATER BIOMES. Various forms of fresh water habitats are seen in this series of photos: a lake, a water garden (pond), and a stream.

Figure 4.52: OCEAN BIOME. Water provides oxygen, food, and recreation.

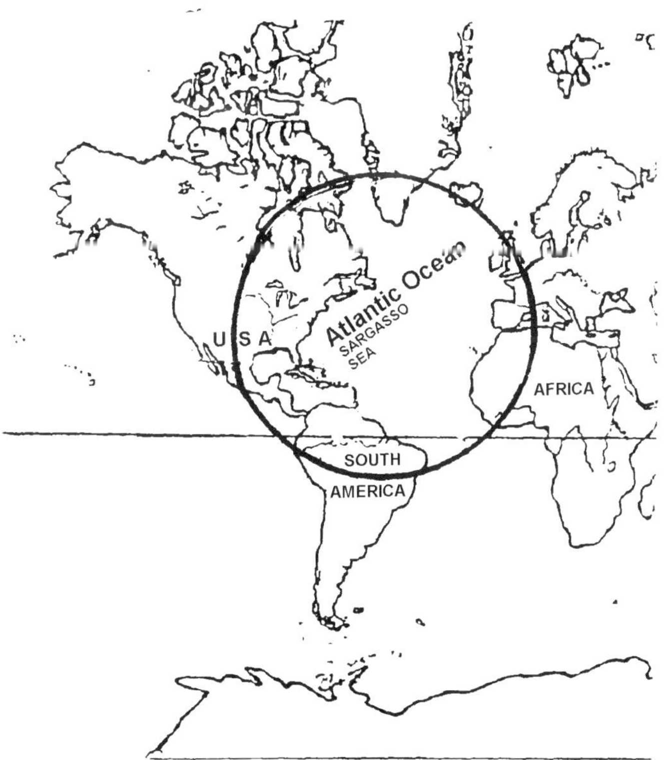

Figure 4.53: SARGASSO SEA. The Sargasso Sea is located in the South Atlantic Ocean and is the home of miles of floating sargassum and its inhabitants.

BIOMES OF THE BIOSPHERE

Biome	Characteristics	Flora	Fauna
Temperate forest	mild climate; temperature varies from freezing to warm summers; moderate rainfall	*deciduous (leaf shedding) *broadleaf; oak, maple, hickory, birch *evergreens: pine, fir, spruce, redwood	deer, squirrel, beavers, geese, ducks, foxes, bobcats, turkeys
Taiga	long, cold winters; warmer temperatures, during summer, melts frozen land and releases enough moisture	*conifers (evergreens) *flowering plants near streams (willows)	bears, weasels, moose, black bears
Tundra	growing season June–July; frozen permafrost; little water available	dwarfed trees, lichen, mosses, low growing plants: sedges, grasses	flies, mosquitoes, lemmings, mice, caribou, musk ox, few species but an abundance of each; grizzly bears, waterfowl, shore birds
Temperate grasslands	low, but sufficient rainfall to support bunches of grasses, etc.	trees only near water; spreading grasses	grazing animals: cattle, sheep, antelope, prairie dogs, kangaroo (Australia), buffalo, jack rabbits
Chaparral	very dry region, low precipitation, limited areas in Mediterranean, S. Australia, S. California— periodic fires assist in reproduction	low shrubby plants; few low trees, evergreen, shrubs, nicknamed shrub-forest	insects, beetles, mule deer, rabbits, bobcats, rodents, snakes
Savannah	rainfall is high, but not spread evenly, rainy season as well as dry season	no forests; grasses, shrubs, few trees	hoofed animals: elephant, rhinos, gazelle, zebra, also tigers, lions, etc.
Desert	very little rainfall; high day temperatures, low night; plant and animal life exist by absorbing dew	cacti, shrubs; yucca	kangaroo rat, lizards. snakes, birds, insects
Tropical rain forest	no seasons; heavy rainfall, no freezing, constant heat	*aerial plants; orchids, air plants *broadleaf plants *rich, profuse plant life, flowers	insects, birds, snakes, frogs, small mammals, monkeys, leopards, very diverse fauna
Aquatic/Ocean	90% of photosynthesis occurs in waters of Earth; temperature range relatively stable, ranging from ___ degrees to ___ degrees.	plankton, kelp, algae, eelgrass, phytoplankton	each aquatic environment has its own species of fish, snails, water beetles, etc.

Figure 4.54: BIOMES OF THE BIOSPHERE. The above chart outlines and gives characteristics of biomes of our world.

✔ THINGS TO KNOW AND DO:

Define:

*biome

*flora

*fauna

*deciduous

*temperate

*transition zone

*conifers

*permafrost

*pampas

*steppes

MATCH THE FOLLOWING

_________ 1. evergreen	a. leaves that shed
_________ 2. tundra	b. grasslands
_________ 3. deciduous	c. conifers
_________ 4. fires	d. permafrost
_________ 5. big game animals	e. savanna

List the nine major biomes:

1.   6.

2.   7.

3.   8.

4.   9.

5.

➜ MODIFIED TRUE-FALSE: correct underlined word if necessary.

_________ 1.   Biomes are usually identified by the <u>fauna</u> of an area. _______________

_________ 2.   Evergreen is another word for <u>conifers.</u> _______________

_________ 3.   Pine, spruce, and fir are <u>deciduous</u> trees. _______________

_________ 4.   <u>Taiga</u> biomes have a constant cover of snow. _______________

_________ 5.   A layer of frozen subsoil is called <u>permafrost.</u> _______________

_________ 6.   Sheep and cattle can be found in the <u>rain forest.</u> _______________

_________ 7.   The <u>Gobi Desert</u> is the largest desert in the world. _______________

_________ 8.   Mammals in the desert get their water from water holes. _______________

_________ 9.   Epiphytes are <u>aerial</u> plants. _______________

_________ 10.   African violets need <u>high</u> levels of light. _______________

✍ FILL IN THE BLANKS:

1.   Most animals in the rain forest live _______________.

2.   _______________ are the most populated places on Earth.

3.   The _______________ zone is along the shores of oceans.

4.   The ocean's basic link to a food web is _______________.

5. The deepest parts of the ocean are called the _________________ zone.

6. Fish are the most numerous organisms living in the _________________ zone.

7. Algae, protozoa, rotifers, and freshwater crustacia are considered _________________

8. In the _________________ zone fish may appear differently, with huge eyes and thin, long tails.

✎ IN SEARCH OF...

Make 3 lists, one of herbivores, one of carnivores, and one of omnivores.

Be thorough: List at least <u>10</u> of each. You may need to use other reference materials.

List 1	List 2	List 3

5

POPULATION

COMMUNITY

ECOLOGY

V. **POPULATION AND COMMUNITY ECOLOGY**

A. Topics that affect populations
  1. Population dynamics and how populations change
     a. density
        1. random
        2. uniform
        3. clumped
     b. biotic potential
     c. environmental resistance
        1. food supply
        2. nesting site
        3. space
        4. water
        5. light
        6. predators
     d. density dependent (biotic)
        1. disease
        2. predators
        3. food supply
     e. density independent (abiotic)
        1. air
        2. water quality
        3. temperature
        4. soil
        5. wind rate
     f. carrying capacity
  2. Community structure and function
     a. community
     b. habitat
        1. species diversity
        2. dominant species
     c. niche
  3. Special relationships among organisms
     a. competition
     b. predation
  4. Symbiotic relationships
     a. mutualism
     b. parasitism
     c. commensalism
  5. Succession
     a. climax community
     b. primary succession
     c. secondary succession
     d. limiting environmental factors
B. Human population
C. Summary

1. Sit in a well-lighted room.
2. Keep distractions to a minimum.
3. Don't recline while studying. You may fall asleep.

Things to Look for in This Chapter:

- Definitions of all highlighted words.
- Differences between primary and secondary succession.
- Differences between a habitat and a niche.
- How populations change.
- Special relationships in competition and predation.
- Three symbiotic relationships.
- Factors that affect the rapid growth of human populations.

TOPICS THAT AFFECT POPULATION

In this chapter we'll present information about populations, behavior patterns within populations, and ways populations are affected by their environment. The ecology of populations may include such topics as:

1. population dynamics: how populations change
2. community structure and function
3. special relationships among organisms
4. symbiotic relationships
5. succession

A **population** is described as a group of organisms of the same species living in the same place during the same time period. A population may be the chickens in old MacDonald's chicken coop, the goldfish in an aquarium, the grass in your backyard, or the ants at the park. The members of a population (blades of grass) may be too numerous for us to count. Scientists take a small sample of the population to study. Actual counts are made from this representative sample and data is collected and interpreted.

Population Dynamics

The actual number of individuals of a species living in a particular area is the **density** of that population. Many abiotic and biotic factors influence the density of populations. Predators, food supply, water, light, temperature, and nesting places are examples of factors that influence density of a population.

You may find in a sample area a great number of robins. Here the nesting material is plentiful; an abundant food supply and low population of predators encourage high numbers of robins. However, in a grassland ecosystem, the necessary requirements for the robin population are not met. Here, there are low numbers of nesting sites, high competition for food, and an abundance of predators resulting in a low density of robins (small population).

Figure 5.1: CHICKENS IN A COOP. The above photo depicts a population of chickens on McDonald's farm.

The density of populations may take three forms. The placement of members of a population within a given area is called **distribution**. There are specific patterns of distribution: random, uniform, and clumped.

1. **Random**: organisms are in no particular order; they are spread out, without pattern, over an entire area.
2. **Uniform:** organisms are spread out in a uniform manner, no more or no less in any one area.
3. **Clumped:** organisms are grouped at points throughout an area; clumping may offer protection from weather/predators.

Populations change in many ways. They may become larger or smaller; they may have more males and less females; even the space they occupy may be increased or decreased. Populations change due to abiotic and biotic factors.

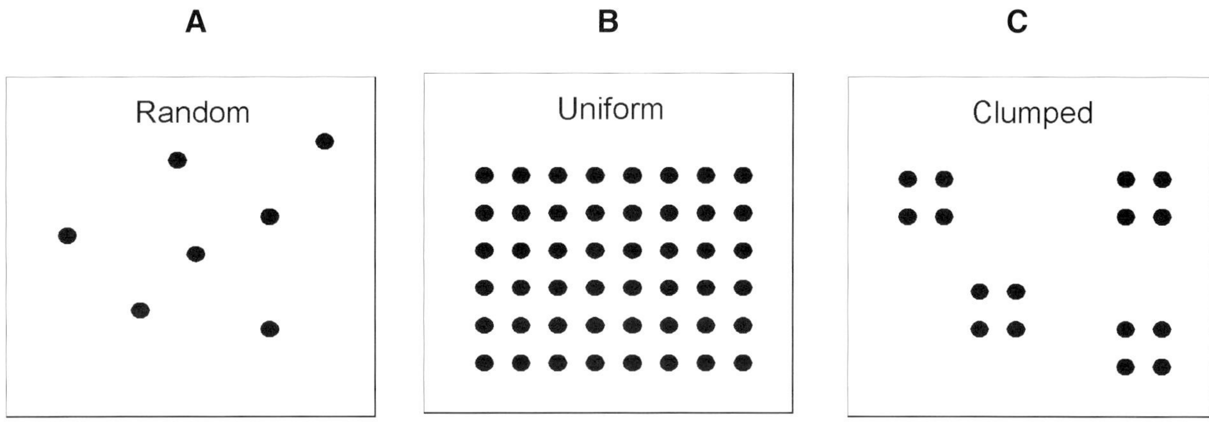

Figure 5.2: DISTRIBUTION PATTERNS. In graphic A, random distribution, members of a population are randomly spread out through an area. In graphic B, members of a population are spread uniformly throughout an area. Graphic C shows populations grouped in clumps throughout an area.

Populations are influenced by reproduction. In the reproductive cycle more offspring are produced than are necessary for that species' survival. Under optimal conditions the maximum reproductive rate is achieved. This is called **biotic potential.** Biotic potential assumes that all of the offspring of any species will survive to their reproductive age. Both abiotic and biotic factors would have to be ideal for this to happen. This is not the case; populations do not reach their biotic poten-tial. Biotic potential is not reached in any species because of **environmental resistance**, those abiotic and biotic factors that limit an organism's survival. Some examples of environmental resistance factors would be the **availability** of the following: food supply, nesting site, space, water, light, predators, etc. A natural balance exists between the biotic potential of organisms and environmental resistance.

Figure 5.3: POPULATIONS. In left photo, we see clumping distribution of both flora and fauna. The right photo shows a random distribution of wildlife.

⬊ DID YOU KNOW?

Some bacteria may reproduce once every 20 minutes. Let's stop and think of what this means at the end of an 8 to 5 workday. A bacterium lands on you at 8 A.M. and begins reproducing.

Time	Amount	Time	Amount
8:00 A.M.	1	2:00	262,144
8:20	2	2:20	524,288
8:40	4	2:40	1,048,576
9:00	8	3:00	2,097,152
9:20	16	3:20	4,194,304
9:40	32	3:40	8,388,608
10:00	64	4:00	16,777,216
10:20	128	4:20	33,554,432
10:40	256	4:40	67,108,864
11:00	512	5:00	134,217,728
11:20	1,024		
11:40	2,048		
12:00	4,096		
12:20	8,192		
12:40	16,384		
1:00	32,768		
1:20	65,536		
1:40	131,072		

Under optimum growth conditions, this **one** bacteria cell has had the biotic potential to produce over 134 million cells by the end of the work day. What are the factors that keep bacteria populations in check?

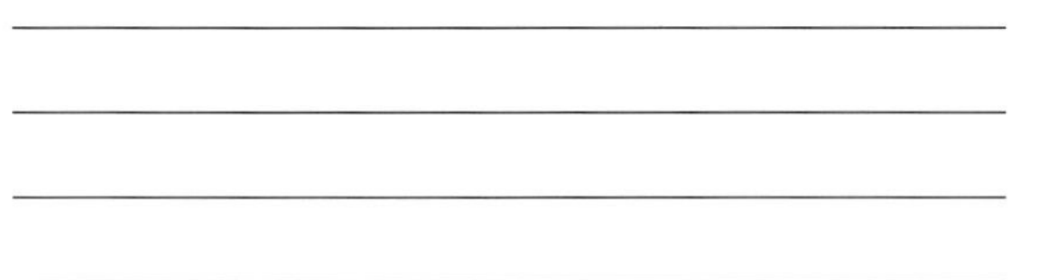

Figure 5.4: ENVIRONMENTAL RESISTANCE. The clearing of land for human use lowers the availability of food supply, nesting sites, space, water, and predators in an area. All the above reduces an organism's chances for survival.

If it were not for environmental resistance, organisms that reproduce within a very short period of time would be out of control. Bacteria, for example, may reproduce every 20 minutes. At this rate, in one day's time, enough bacteria would be produced to cover the entire Earth's surface over a half foot deep. Environmental resistance is responsible for maintaining a balance between the environment and population numbers. This balance is delicate. It is, however, a working one and continues to be, when not influenced by man.

The factors that influence populations may be termed **density dependent** or **density independent**. **Density dependent factors** are usually biotic: disease, predators, and food supply. These factors will influence a crowded area differently than a sparsely-populated one. Disease, for example, spreads very rapidly in a crowded area, whereas in less populated areas, the rate is slower. Another density dependent factor, a shortage in the food supply, would be more devastating in a highly-populated area than in a less-populated area.

Abiotic factors are usually associated with **density independent factors**: air, water quality, temperature, soil, and wind rate. These factors influence all organisms with density being no issue. For example, a grass fire on the prairie affects all organisms, without regard to population number. The fire destroys the food supply and shelters of all organisms whether they are in groups or scattered throughout the area.

Population dynamics is also concerned with **carrying capacity.** Carrying capacity represents the number of individuals an area can support in terms of food, shelter, and space. Populations in nature usually function at their carrying capacity. **Environmental resistance** and **biotic potential** together keep most populations stable and hinder them from exceeding their carrying capacity. A plot of land can only support so many organisms. Plants and other living organisms depend upon nutrients in the soil, the available water, food, and shelter. With all these factors, nature still maintains a balance.

Man has negatively influenced the carrying capacity by altering the environment to suit his needs. He has developed land, causing populations to lose some of their space. This remaining space now becomes more densely populated increasing the demand for food, shelter, and water. In this new situation, the popula-

Density Dependent Factors
1. disease
2. predators
3. food supply

Density Independent Factors
1. air
2. water quality
3. temperature
4. soil

Figure 5.5: DENSITY DEPENDENT/INDEPENDENT FACTORS. Abiotic and biotic factors may be either density dependent or density independent. Both groups exert their influence on organisms in any environment.

tions must decrease and a new balance must be achieved. In what ways could this happen?

____

____

____

____

Figure 5.6: CONDO ON THE BEACH. Humans negatively influence the carrying capacity by altering the environment to suit their needs. Over development of our coast line has reduced population numbers of native species.

Figure 5.7: POPULATIONS AFFECTED BY MAN. The development of our shoreline has reduced the populations of sea oats, various marine life, and even sand dunes. Right photo is a nesting Loggerhead sea turtle whose nests are now protected along Florida beaches. (Ron Mezich, Florida Department of Environmental Protection).

↘ DID YOU KNOW?

As of 1996 China has the highest population in the world. There are 1.17 billion people in China, more than 1/5th of the world's population.

✔ THINGS TO KNOW AND DO:

Define:

* population ____

____

____

*density ____

____

____

*random distribution ____

____

____

*uniform distribution ____

____

*clumped distribution ____

____

* biotic potential ____

____

* environmental resistance ____

____

*density dependent ____

____

*density independent ____

____

*carrying capacity ____

____

List the abiotic factors that are associated with density independent factor:

1. _______________________   4. _______________________

2. _______________________   5. _______________________

3. _______________________

→ **MODIFIED TRUE-FALSE: correct underlined word if necessary.**

________ 1.   Biotic potential assumes <u>not</u> all of the offspring survive to their reproductive age. _______________________

________ 2.   Biotic potential is <u>not</u> reached in any species because of environmental resistance. _______________________

________ 3.   A natural balance <u>exists</u> between biotic potential and environmental resistance. _______________________

________ 4.   Food supply, nesting site, space, and water <u>are</u> some of the environmental resistance factors. _______________________

________ 5.   Some bacteria are able to reproduce every <u>5</u> minutes. _______________________

✍ FILL IN THE BLANKS:

1. _______________, _______________, and _______________ are density dependent factors.

2. In populated areas _______________ spreads quickly.

3. _______________ destroys food supply and shelters of all organisms whether they are scattered or not.

4. A plot of land can only _______________ so many organisms.

5. Man has _______________ the carrying capacity by altering the environment to suit his needs.

IDENTIFY:

1. Label the following:

_______________ _______________ _______________

2. The density of populations may take 3 forms; name them.

_______________ _______________ _______________

3. What types of factors influence the density of populations? List 2 from each group.

A. _______________________________ B. _______________________________

    1. _______________________   1. _______________________

    2. _______________________   2. _______________________

4. Give **three** requirements that may or may not have been met for the following populations:

a. an area with high robin population:

b. an area where the rabbit population is low:

✎ IN SEARCH OF ...

How would you count the number of blades of grass in your back yard?

____

____

____

Count the blades of grass in your back yard. How many are there? _______________

Community Structure and Function

In nature a given space is seldom occupied by a single population. Observation would show numerous flora and fauna. **Communities** consist of different populations living within the same area. Populations in a community are dynamic, changing, and dependent upon one another.

Different ecosystems have different flora and fauna. The frozen polar region supports its own specie types as do the ocean ecosystems. **Species diversity** is the varying of species from ecosystem to ecosystem.

In each ecosystem there are one or more populations that are more important than the other species; these are called **dominant species.** A dominant specie strongly influences the food supply and the environment of other species within that ecosystem. The West Coast of the United States is home to a dominant tree—the redwood.

The maple is the dominant tree on the East Coast of the USA. This tree provides a canopy which regulates the amount of light that reaches the understory and forest floor. Much

Figure 5.8: POLAR BEAR. In our northern climates there is great specie diversity. This polar bear represents one dominant species in this environment.

of the litter that helps build soil and provide food comes from fallen leaves. Maple trees provide a place of attachment for certain species as well as shelter for others. Mosses on the tree bark and birds' nests are examples of provisions provided by this dominant species.

The importance of the dominant species is tremendous. As mentioned, it exerts its influence in so many ways and areas of an ecosystem. Every organism, no matter how big or small, has to have a place to live. This special living space within a community is called an organism's **habitat**.

Organisms need certain habitat requirements; however, some organisms may be able to live in a variety of places or in just one. Rats are able to live in many different places. We see them in the city, country, near aquatic environments and within homes. (Of course, not yours or mine.) They have **general**, rather than **specific**, requirements; therefore, they exist in a variety of habitats.

Other organisms, such as the ghost crab, have very specific requirements for their living space; therefore, they occupy a specific habitat, a sandy beach. Within the organism's

Figure 5.9: PALM TREES. Palm trees represent dominant species in very warm climates, providing food and shelter for species living within that community.

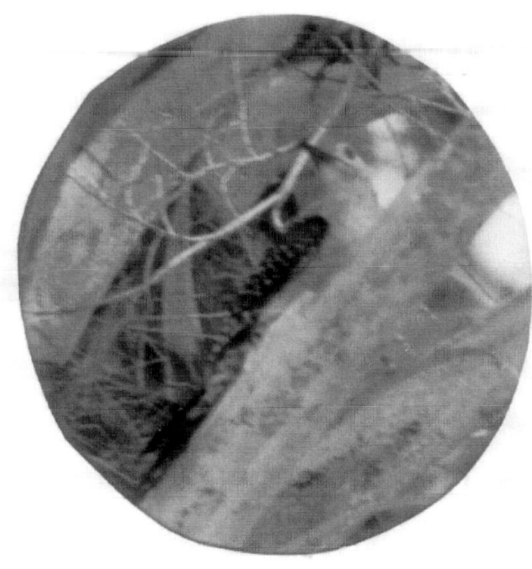

Figure 5.10: SHELTER. This palm tree provides shelter, protection, and food for a variety of organisms such as the bird in the above photo. (B. Schmidt)

Figure 5.11: HABITAT. This chameleon's habitat is the area within and around the limbs of this palm tree. (B. Schmidt)

habitat are all the specific requirements, both abiotic and biotic, that the organism needs.

The particular role or lifestyle of a specific population in a habitat is called a **niche**. Some plants and animals share the same habitats in the ecosystem. Even if they share the same habitat, they occupy a different ecological niche.

Insects, for example, live in the same habitat, but each occupy a different niche. The grasshopper, being active during the day, does not interfere with the cricket living in the same habitat who is active at night. Both insects occupy a different niche in the same habitat. One feeds on the leaf tops while the sun is shining, and the other occupies the space beneath the fallen leaves. At night the crickets come out from under the leaves, feeding and filling the area with song. In review, the habitat is the spot or location where the organism lives, and the niche is its particular role within the community.

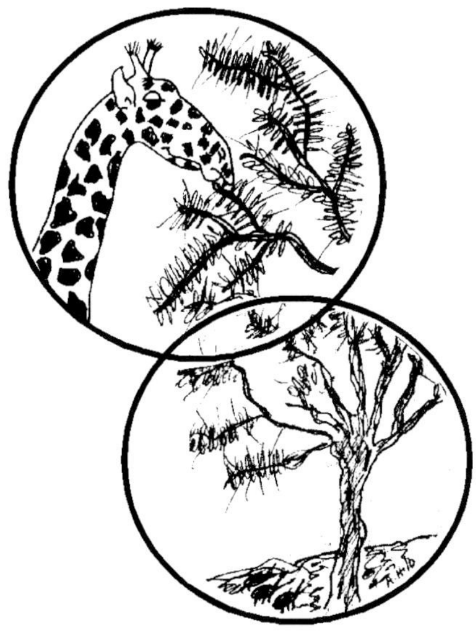

Figure 5.12: NICHE. The particular role or lifestyle of a specific population in a habitat is called a niche. In this graphic, both the giraffe and the insects are living in the same habitat but occupy a different niche. The giraffe feeds by day on the acacia tree and insects by night.

✔ THINGS TO KNOW AND DO:

Define:

* community ____

____

____

* species diversity ____

____

____

*dominant species ____

____

____

*habitat ____

____

____

* niche ____

____

____

List the dominant species in the following ecosystems:

* a forest in California. ____

* a forest in Massachusetts. ____

* New York City ____

* a gold fish tank ____

____

____

Name two organisms that can live in a variety of places.

1. _______________________________   2._______________________________

Name two organisms that have very **specific** habitat requirements.

1. _______________________________   2._______________________________

✎ FILL IN THE BLANKS:

1. A dominant species _____________________ __________________the food supply
    and the environment of the other species.

2. Even if some organisms share the same _____________________, they may occupy a dif-
    ferent _____________________.

Special Relationships among Organisms

The interaction between various organisms with an ecosystem may or may not be very specific. The types of interaction depend upon the habitat and species of populations within that habitat. Organisms may or may not interfere with each other in terms of feeding requirements, territorial needs, or water supply.

When the same food is used by more than one species, or even an abundance of the same species, competition comes into play. **Competition** is the act of striving for food, water, shelter, mates, etc. by a number of organisms located in the same habitat.

Habitats have a limit to the amount of food, water, shelter, space and light, and minerals available. Competition for one or more of these factors becomes an important issue for the organism's survival. The closer one population's requirements are to another, the greater the level of competition.

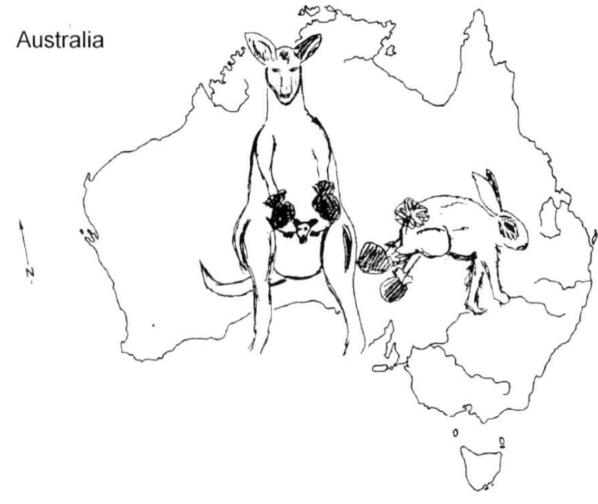

Figure 5.13: COMPETITION. When food, territorial needs, or water supply is used by more than one species within a habitat, competition comes into play. Animal survival may be greatly affected by the level of competition.

Figure 5.14: PREDATION. The eagle is a very capable predator. With its strong curved beak and grasping talons, it is able to provide itself with nutrition. (U.S. Soil Conservation Service)

Direct intense competition between organisms may lead to the extinction or adaptation of one of the competitors. If the organism is not able to adapt in some way to this level of competition it perishes. An organism may adapt by being able to survive in another part of the environment, feed on a different food source, or even feed at different times of the day. Whatever the case may be the organism survives, reproduces and remains alive.

Competition may be reduced when organisms establish their own territories and stay within them. Darwin, as he studied the Galapagos Islands, noticed many types of finches living together. Each type had adapted to different niches within the same habitat. Some of the birds had long, pointed beaks for obtaining insects in tree bark crevices. Others had strong, thick beaks for crushing seeds. Many other variations were noticed by Darwin as he observed the finches. Each specie variation had a particular niche, with no overlapping; therefore, the competition level was low.

Another type of relationship occurs when organisms hunt and kill their prey. This relationship is called **predation**. Even though predators look very different than one another, they share similar characteristics. They have a keen sense of smell, sight and hearing. Usually they are very quick and nimble. All these characteristics help predators capture their prey.

High numbers of prey within a habitat increase the predator's food supply. When an area is overrun with an abundance of prey, the prey have an increased chance of becoming sickly and malnourished. This intense competition level for food takes its toll on the prey. The prey in this weakened state are more easily accessible to predators. Predators reducing overpopulated areas (density dependent) help keep nature's balance intact.

The prey, on the other hand, have also adapted to the situation. Distinct characteristics have aided them in the prevention of their capture. Their senses are finely tuned. Sharp hearing, a keen sense of smell, and coloration assist them in their survival.

Figure 5.15: CAMOUFLAGE. The coloration of this rabbit blends in so closely with the environment that he is hardly noticeable—to us as well as his predators. His camouflage helps him stay alive longer.

Figure 5.16: CAMOUFLAGE. A deer is hardly noticeable while standing still, hidden by his camouflage coat.

Some organisms blend in with their environment which helps them stay alive. **Camouflage** can be found in organisms in every habitat in every part of the biosphere. The familiar chameleon changes its color to match and blend in with its surroundings as do octopi in the marine environment. Unchanging coloration can also conceal an organism from its prey with great success. The coloration of a flounder is light on the bottom and dark on the top. As it lies on the sandy bottom, it is so well camouflaged that an onlooker usually cannot detect it from the sand. A fish that might prey on a flounder will only see light if the flounder is swimming above the predator; it will only see sand if the flounder is below. The flounder has perfect coloration for its environment. Another example of camouflage is the heron. A sole heron standing along the river bank with the vegetation behind him is nearly invisible. His coloration protects him from predators and conceals him from the prey which he is stalking. As an organism is more suited for its environment, the greater the chance is for its survival.

↘ DID YOU KNOW?

...That flounder bury themselves in the sandy bottom of the ocean or gulf when they sleep or that they have both eyeballs on one side of their face? Check this out at your local fish market.

✔ THINGS TO KNOW AND DO:

Define:

* competition ____

____

____

* predation ____

____

____

* camouflage ____

____

____

List 4 items that organisms compete for:

1.____

2.____

3.____

4.____

Describe 6 factors that are limited in an organism's habitat.

1. ______________________   4.______________________

2. ______________________   5.______________________

3. ______________________   6.______________________

✍ FILL IN THE BLANKS:

1. The ______________one population's requirements are to another, the

______________the level of competition.

2. ______________ ______________competition may lead to

______________or ______________of one of the competitors.

3. Predators have a ___________________ sense of smell.

4. Chameleons and octopus use ___________________ to protect themselves.

5. The greater the ___________________ ___________________, the greater the

    ___________________'s chances to capture its ___________________.

Symbiotic Relationships

Another set of relationships to study are linked to symbiosis. **Symbiosis** occurs when organisms live closely together in a specific relationship. When two organisms live together and each benefits from the relationship, this is called **mutualism.** One common example of mutualism is found in lichens. Lichens are algae and fungi living together, both benefiting from the relationship. Algae produce food through the process of photosynthesis, providing food for the fungi. The fungi provide water, minerals, and a place of attachment for the algae. This is only one mutualistic relationship; there are many more in nature. Can you think of any?

____

____

Organisms on the land and in the ocean are involved in this type of symbiosis. Another example of mutualism is seen between the rhinoceros and the oxpecker bird. The oxpecker feeds on the pests that infest the rhino's skin, and it, in turn, receives protection from the rhino. The oxpecker cleans the rhino's skin and acts as an early warning system of approaching danger. Another mutualistic example is the remora fish and other marine organisms. Remora fish in the ocean attach themselves to sharks, rays, large fish, sea turtles, and whales. The remora receives a free ride and also eats food scraps left by the host (sharks, etc). The host, in turn, rids itself of parasites.

Figure 5.17: LICHEN ON FOREST FLOOR. Lichens are an example of a symbiotic relationship called mutualism. Here algae and fungi live together, both benefitting from this relationship.

SYMBIOSIS

1. **MUTUALISM**
    - oxpecker/rhino
    - lichen (algae and fungi)
2. **PARASITISM**
    - lice/people, cows
    - tapeworm/people, dogs
3. **COMMENSALISM**
    - innkeeper worm/crabs, clams
    - roosting birds/trees

Figure 5.18: SYMBIOTIC RELATIONSHIPS. The above chart gives various examples of symbiotic relationships.

Figure 5.19: MUTUALISM. Remora fish attach themselves to various marine organisms. The remora fish receives transportation and food supply left by the host; the host rids itself of parasites through the action of the remora fish.

Parasitism is another common symbiotic relationship where one organism, the parasite, lives off of another organism, the host, for much or all of its life. The parasite depends on the host for food. It may live internally or externally on the host. The host is harmed in some way by this relationship. The weakened host may die, at which time the parasite has to find a new host or it, too, will perish.

Humans have parasites which include tapeworms, lice, ticks, leeches, and a wide range of others. Lice are usually associated with the hair of school children. Children are in close contact with one another, playing, touching, exchanging sweaters and hats. Lice are carried along with these items and before long many students have "itchy heads." Schools do not allow students back into class until their heads are checked by the school nurse or health department. Special shampoos kill the lice and their eggs that have been laid. The egg cases are tightly attached to the hair shaft, so removing them is no easy job.

Tapeworms, another parasite, usually enter the body through uncooked and improperly cooked foods. The tapeworm attaches itself to the wall of the intestine by hooks and suckers found on its head. Tapeworms absorb the nutrients and may cause the host to become nutritionally deficient. There are medicines that can be taken to get rid of tapeworms. There are many wise tales and jokes about tapeworms. One of our favorite jokes came from Grandmother Martino (not original, I'm sure).

Two men sat down for lunch at a construction site. They pulled out their lunches and began eating. One man said, "Boy that smells like rotten eggs." The other man replied, "That's just what it is. I have a tapeworm and this is good enough for him."

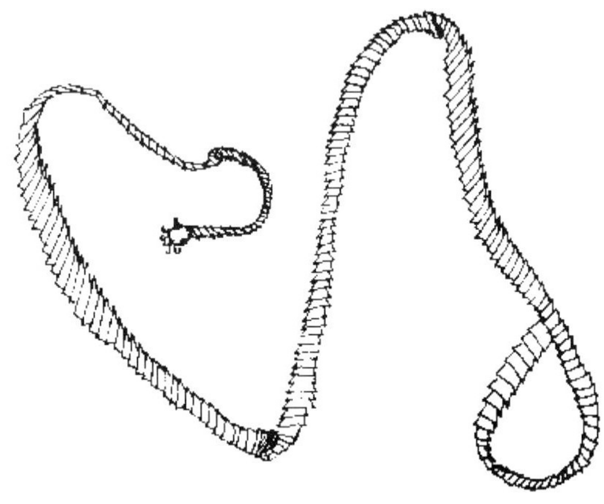

Figure 5.20: TAPEWORM. The tapeworm exhibits a parasitic relationship between itself and its host, which may be numerous organisms.

Parasites are found all over the Earth. We have joked about them, but they need to be taken seriously. Sanitation and proper hygiene are methods that cut down infestation and minimize this symbiotic relationship.

In review, parasitism is a symbiotic relationship where one organism benefits and one organism is harmed or perishes; mutualism is a symbiotic relationship where both organisms benefit, and the third symbiotic relationship is **commensalism** where one organism benefits and the other is neither helped nor harmed. There is a worm in the ocean called an innkeeper worm that lives within a burrow it has created. Other organisms, clams, worms, crabs, small fish, etc. take up residence in its burrow. The innkeeper worm creates water currents which sweep food particles into the burrow. There is an ample supply of food to feed the worm and all the other inhabitants of the innkeeper's burrow. The worm is neither hurt nor helped by their

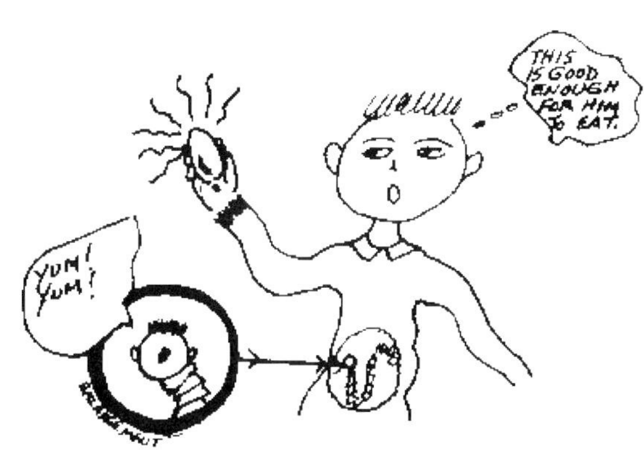

Figure 5.21: SUPPER TIME. "I have to feed my tapeworm. He won't mind the rotten egg."

presence. The other sea creatures receive protection and food in this relationship. Symbiotic relationships are a part of nature's delicate balance. Can you think of a symbiotic relationship where you live?

____

____

____

____

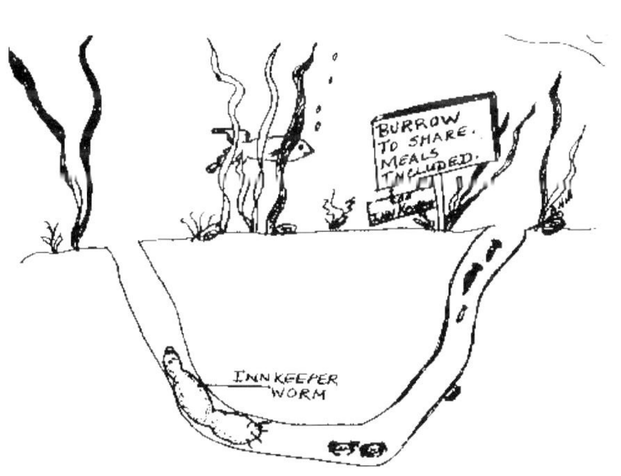

Figure 5.22: COMMENSALISM. Innkeeper worm shares his home and food supply with others. He receives nothing in return.

↘ **DID YOU KNOW?**

What relationship exists between legumes and the nitrogen-fixing bacteria that live within the nodules on their roots?

✔ **THINGS TO KNOW AND DO:**

Define:

*symbiosis ____

____

____

*mutualism ____

____

____

*parasitism ____

____

____

*commensalism ____

____

____

Give an example of the following symbiotic relationships and describe in detail the role each partner plays in the relationship.

mutualism:

____

____

____

parasitism :

____

____

____

commensalism:

____

____

____

Succession

Other relationships in nature exist due to succession. Succession can be seen in barren fields, land burned by a careless hunter, or cleared land left idle. The orderly replacing of various communities one after another in an area is **ecological succession. Succession** is usually gradual, taking place over a fair amount of time. To the land owner who clears his fields, succession takes place too quickly. He sees the tall grasses and small shrubs taking over his cleared field. But the events of succession are quite specific in a given biome. Each specie has its place in the succession process like the tall grasses and small shrubs that grow up in a farmer's unused field.

Let's start at the beginning of the process. Abiotic factors such as water, solar energy and nitrates in the soil exert an influence on the populations that survive in an area. Plant populations are most affected by abiotic factors. Ultimately the type of soil (nutrient content and type) and the climate (moisture and temperature) are the two abiotic factors that control where plants survive. Plants, being the food source for animals, determine which animal population will be present in an ecosystem.

As well as the abiotic factors influencing population, biotic factors also exert influence

Figure 5.23: FARM. This was once a plowed field. Now, left abandoned, this field is overgrown with grasses and shrubs, signs of succession.

on the non-living environment. For example, soil structure, water-holding capacity, and acidity (all abiotic) are influenced by the decaying litter (biotic) on the forest floor.

Interacting biotic and abiotic factors produce changes in the ecosystem. Over time, conditions change, and populations come and go due to their ability to survive in a specific area. Ecologists have studied succession in ecosystems over a time period. They have seen the successive replacement of one community with another. As mentioned, succession is orderly and predictable in each biome. The out-

Figure 5.24: DRY-LAKE BED. Looking closely to below the tree level, pick out the old lake pier. Water level in this area dropped over time, exposing the lake bed. Succession proceeded along, resulting in grasses and other small plants invading the area.

come of succession leads to a specific community structure that is relatively stable.

Stable ecosystems undergo little further change. When a stable structure is attained within an ecosystem, it is called a **climax community.** In reality, we understand that change, even to a tiny degree, is constant, always happening.

To our eye the ecosystem appears very stable, unchanging, but even the most stable ecosystem is always changing. For example, one of the most stable terrestrial ecosystems is the redwood forests of California. This climax community is a few thousand years old, but very young when considering the age of our planet.

Succession can best be monitored in situations where a place has been disturbed by one of the following:

1. Climatic changes (warming or cooling of an area),
2. Geological occurrences (rock slides, volcanic eruptions), and
3. a large-scale human action (clearing of a forest, cultivation for agriculture, and subsequent abandonment).

Figure 5.25: LITTER. Fallen leaves, branches, and other plant and animal remains cover the forest floor. Litter influences soil structure, water-holding capacity, and acid level of an area.

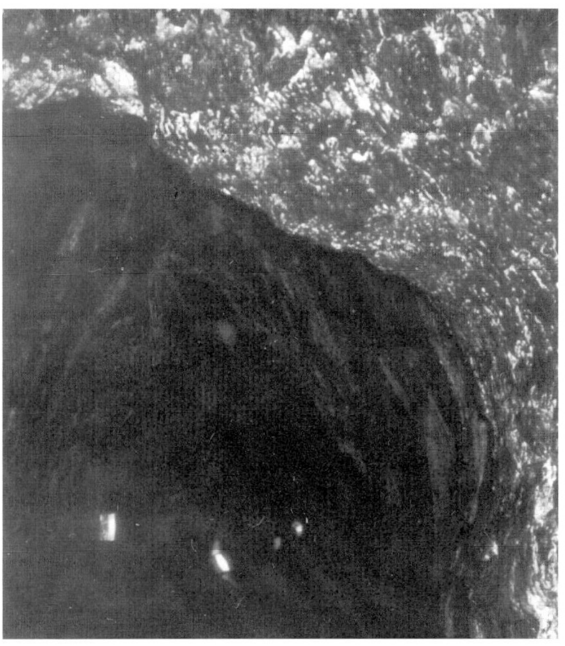

Figure 5.26: HAWAII VOLCANO NATIONAL PARK. In the left photo, you are looking at a climax community that once appeared like the right photo. Volcanic action created a lava tube, bared of all living organisms. Through the process of succession, a climax community has resulted.

Primary Succession

Primary succession occurs when a potentially habitable ecosystem is available for colonization. This ecosystem is essentially lifeless and usually devoid of soil. The receding of glaciers in the Northern United States about 10,000 years ago developed just such a situation. As the glaciers melted, a soil-less area of bare rock was left. Today the same chain of events is developing in Alaska at Great Glacier Bay. Biologists have been observing the entrance of new communities there as the glaciers have been melting and exposing the bare rock for the last 200 years.

Another candidate for primary succession was the severe earthquake that struck Madison River Canyon in Southern Montana near Yellowstone Park on August 17, 1959. A huge rock slide came crashing down into Madison Canyon. It has been estimated that enough rock fell to fill the Rose Bowl, a "heaping" ten times over. The ecosystem of this area was immediately and completely changed. Soil, plants, and animals were completely wiped out. Years passed after this rock slide event and the soil rebuilding process began. The first

Figure 5.27: LAVA FLOW. The above photo shows an area devoid of all soil, plant, and animal life. Primary succession will occur here, taking many years to complete.

Figure 5.28: SUCCESSION ON A LAVA FLOW. Refer back to the previous photo. This photo represents succession on a lava flow after 110 years.

organisms to come in were lichens, organisms made up of algae and fungi; they are called **pioneer organisms.** Pioneers act on the environment along with wind, rain, ice, and temperature. The rocks begin to break into soil. Lichens help to trap soil being blown by the wind and transported by water. As the soil forms and new organisms settle in the area, lichens die. Mosses, grasses, and small flowering flora settle in, grow, and reproduce. The first organisms to settle in an area such as this must be drought-resistant plants. New species enter the area preparing the way for stiii other types of organisms. These events are repeated, each time adding some slight change to the ecosystem. After hundreds and hundreds of years of succession, a climax forest may result if the area is left untouched. Primary succession may therefore be defined as "succession from bare rock."

Figure 5.29: MOUNT ST. HELEN'S VOLCANO, MAY 18, 1980. Shock waves leveled fir trees for 70 square miles. The top of the mountain was blown away. (Mt. St. Helen's-2. U.S. Geological Survey)

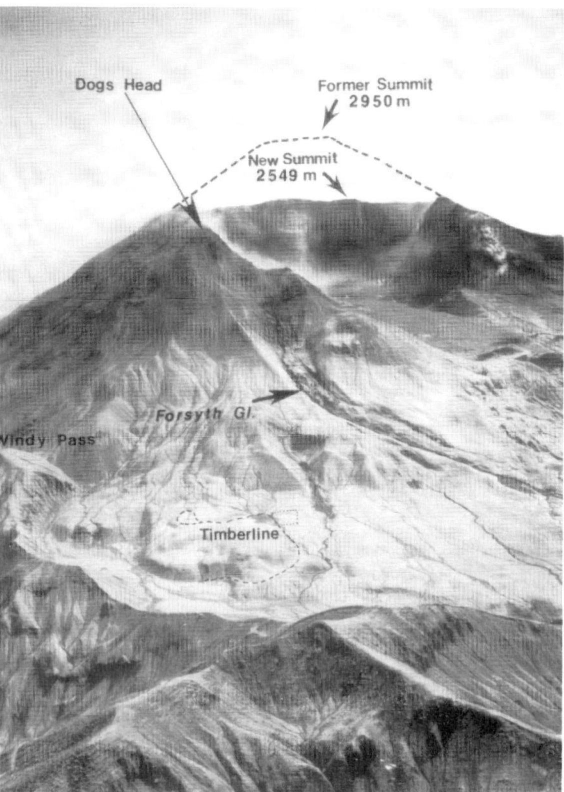

Figure 5.30: MOUNT ST. HELEN'S. The above photos show the mountain top before and after the volcanic eruption of Mt. St. Helen's. Notice the figures at the top of the second photo. (MSH-Brugman, M.M, 1. U.S. Geological Survey)

SECONDARY SUCCESSION

Most succession is not as drastic as the preceding examples. Clearing land, severe storms like Hurricane Andrew that hit Southern Florida in 1993, and forest fires provide good areas for studying succession. In these situations, the soil has not been completely removed and succession will happen at a faster rate. Succession in an area following destruction of an existing community by fire or storm is called **secondary succession.**

Secondary succession can be seen over a short time span. The time frame difference between secondary and primary succession is noteworthy. The presence of soil in secondary succession allows for a faster succession. The soil contains organic plant material. All these items help increase the speed of succession.

In the United States, long before European settlers arrived, beech, maple, and hickory trees covered much of the land. People settled this land first, by clearing fields and next, by leaving and clearing new fields. The abandoned fields changed from farm land to a weeded area. Grasses, shrubs, conifers (pines), and finally deciduous trees (oak, birch, etc.) took over the abandoned fields, one right after the other. This resulted in a stable **climax community**.

Figure 5.31: HURRICANE DAMAGE. Inland storm damage uprooted many trees and eroded river banks. Unfortunate as this is to our environment, it is a prime place for us to study secondary succession.

The climax community in all succession depends upon the particular environment at hand. For example, if water is limited, the area may stabilize as a grassland or a scrub forest. Some environmental factors that influence succession are:

1. soil condition
2. drainage
3. water supply
4. type of seeds transported into the area
5. type of community that previously existed

Other succession takes place as ponds and lakes age. Young ponds have clear water, practically devoid of flora, fauna, and nutrients. As time passes nutrients filter into the water mainly by runoffs from the land. Algae grow well in this situation. Plants (rushes, sedges) grow along the edges of the pond and later die. Their remains fall into the water, decaying and littering the pond floor. Plants and animals in the water grow, reproduce, and die. More matter is added to the pond's floor, making the lake more shallow. Land plants growing along the edge of the pond become more numerous.

Figure 5.32: SUCCESSION IN A POND. Young ponds have clear water and distinctive borders. As ponds age, water becomes cloudy, bottoms become littered, and borders become varied and vague, and even appear smaller.

The pond now looks smaller. Eventually enough soil has been deposited into the pond by the wind and it becomes shallower. Aquatic plants die off from the lack of abundant water and new species grow in their place. Eventually grasses, woody shrubs, and trees may occupy the area where the pond once thrived. These new populations become the climax community.

To sum it up, succession depends upon the "invading" populations and the environmental factors present in a specific area.

↘ DID YOU KNOW?

One spot on the Earth where primary succession is occurring is on the Island of Hawaii. Hot-lava flows, reaching the cool ocean water, explode and become instant beachfront property.

✔ THINGS TO KNOW AND DO:

Define:

*ecological succession ____

____

____

*climax community ____

____

____

*stable community ____

____

____

*primary succession ____

____

____

*pioneer organism ____

____

____

*secondary succession ____

____

____

*invading populations ____

____

→ MODIFIED TRUE-FALSE: correct underlined word if necessary.

________ 1. Succession is usually <u>gradual</u>. ________________

________ 2. Each species has <u>various</u> places in succession. ________________

________ 3. Abiotic factors such as <u>food</u> influence populations that can survive in an area.

____

________ 4. <u>Plants</u> determine what animal populations will be present in an ecosystem.

____

________ 5. Succession is <u>never</u> orderly and predictable in each biome.

____

________ 6. Climax communities change <u>rapidly</u>. ________________

________ 7. Primary succession takes place in an area that is <u>not</u> void of soil.

____

________ 8. Pioneer organisms are usually in the <u>middle</u> stage of succession.

____

________ 9. Lichens are <u>pioneer</u> organisms. ________________

________ 10. <u>Secondary</u> succession is defined as succession from bare rock.

____

✍ FILL IN THE BLANKS:

1. Environmental factors that influence succession are:

    1. ____

    2. ____

    3. ____

    4. ____

    5. ____

2. ________________, ________________ ________________, and
________________ may occupy the area where ponds once thrived.

3. Succession depends upon ________________ ________________ and the
________________ ________________ present in a specific area.

HUMAN POPULATION

The human population in 1990 reached a whopping 5.3 billion people. This high figure is due to its tremendous current growth rate. The growth of the human population is more rapid today than ever before. Today it takes less time for our population to double than it did in 1 AD. The **doubling rate** in 1 AD was 2,000 years. In the year 1000 AD it only took 600 years for the population to double. By the year 1975, the doubling time was only 45 years.

Why is the doubling time decreasing as time goes on? Why is the world's population continuing to grow at such a rapid rate? Birth rate and death rate influence the population growth rate. In the early history of mankind, death due to disease (density dependent), accidents, and poor hygiene kept a close check on the population size.

Early human populations are assumed to have been limited to savannas. Their food source consisted mainly of plant matter with small amounts of meat. As time marched on,

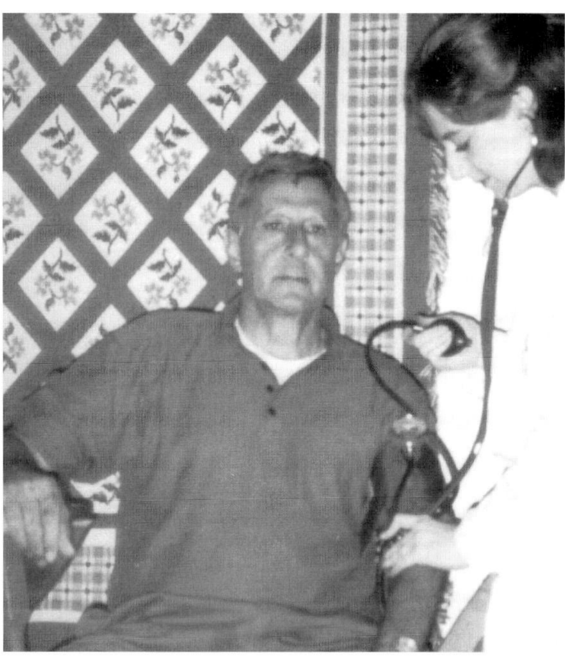

Figure 5.34: HOME HEALTH CARE. Death rate is reduced through programs such as this one where visiting medical professionals bring care to those who are home-bound. How does a lower death rate affect populations?

DOUBLING RATE

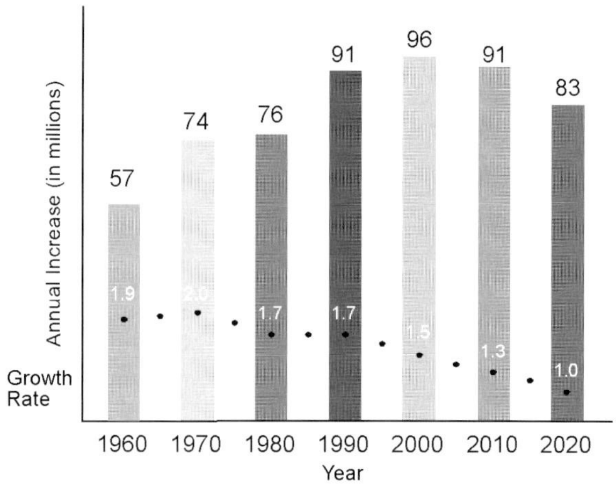

Figure 5.33: POPULATION RATE BY YEARS. Population has increased rapidly in the past. According to projections, by the year 2000, growth should level off and begin to decrease. (Biosphere 2000).

so did man. Hunters spread out and gathered food where they could find it , thereby increasing the area populated by man.

Most animal populations do not spread out very far or as rapidly as mankind. Because of man's intellectual abilities, his building of shelters, his use of fire, and his use of tools, man's control of the environment across the Earth was inevitable. Turning towards an agricultural way of life, domesticating animals, and irrigating fields, men began to form communities. Primitive farms provided a more dependable food supply. This allowed the human population to grow. The increased food supply also increased the carrying capacity for the human population in these settled agricultural areas.

During man's early days, we did not have the degree of control over the environment as we do today. We were subject to the same population-limiting factors as the rest of Earth's organisms. Of course, we always had the ability to manipulate our environment to some degree.

Technological advances in agriculture, construction, communication, and transportation have made it possible for us to live almost anywhere at the present time. Today we are even reaching out and searching for places to live other than planet Earth. We have gained knowledge in the health fields that keeps us living longer. If we are attacked by disease or have an accident, we have the medicines, doctors, and hospitals to help us recover. This improved health care, along with our own knowledge of how our bodies work and what keeps us healthy, reduces the death rate.

As we take care of those around us who are weak, abnormal, or old, we are keeping the death rate down. In nature the weak do not survive. It is the fastest, strongest, keenest, and best hunters that survive. As more of the population survive to the reproductive age, more offspring are produced. The birth rate is increased because of the increased survivorship of humans.

If the death rate is decreased and the birth rate is increased, the population has nowhere to go but up, up, up! As medical practices and sanitary conditions improved, the population began to grow slowly, at first. Contagious disease (density dependent) still spread rapidly through highly populated areas. Carriers of diseases (rats and fleas) allowed disease to spread. This kept the population down.

You may recall from health or history classes that Black Death wiped out two-third's of Europe's population between 1347 and 1351. Now, with widespread use of vaccines,

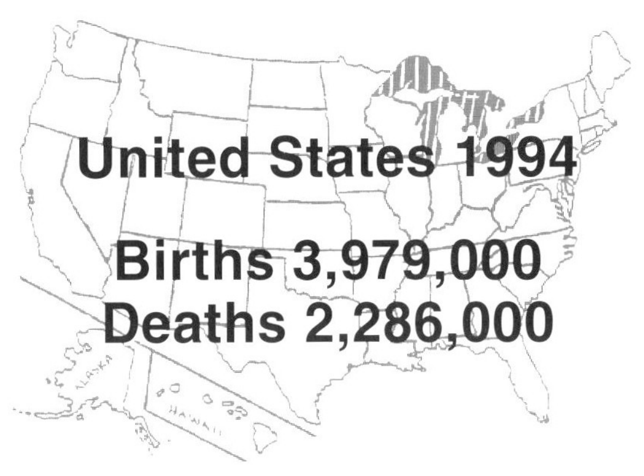

Figure 5.35: U.S. CENSUS STATISTICS FOR 1994. Birth rate is greater than the death rate; therefore, population increases.

antibiotics, and insecticides, these huge epidemics seldom occur. Infectious diseases today are no longer a major obstacle to population growth. However, the AIDS virus (Acquired Immune Deficiency Syndrome) in the world's population has increased the death rate.

Another reason our population continues to grow is man's ability to harness the energy of fossil fuels such as coal; this led us down the path to an even greater growth rate. All these discoveries removed some of the factors that had once checked our population.

This tremendous brain power of mankind has gotten us into somewhat of a serious predicament. Our demand for food is always on the upswing. We are striving to produce better and better crops and livestock. We need crops that produce great yields, crops that are very resistant to cold, heat, drought, insects, and other crop-damaging organisms. We breed farm animals that produce more milk and more meat and larger eggs while being less susceptible to diseases. However, with all of our intelligence and knowledge, we still have people malnourished, starving, and without adequate shelter.

The figures are staggering as we take the time to pay attention to studies done in this area world wide. We find that approximately one out of six people are starving, without shelter, and without healthcare. In fact, about

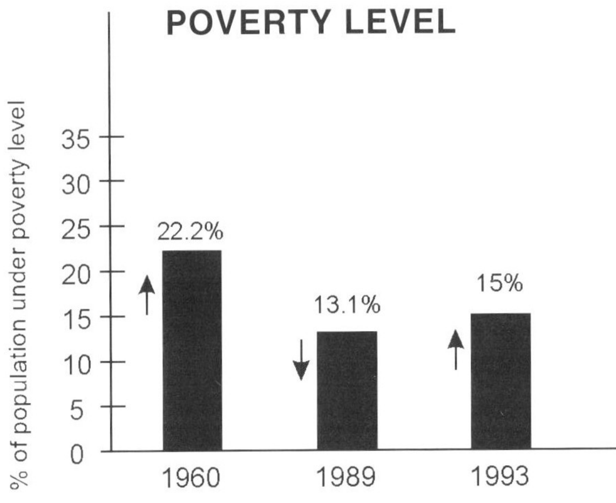

Figure 5.36: POVERTY LEVEL IN U.S. According to 1996 statistics, the poverty level once again is on the upswing. How does today's population influence our poverty level?

one-third of the Earth's population has no healthcare or sewage treatment facilities. At the rate we are growing, the Earth cannot sustain us.

We must look into the human population situation "today." It is up to you and me to come up with solutions now! For example, today teenage pregnancies are increasing because of several factors:

- lower moral standards
- too much independence too soon
- lack of self-control
- decay of families

As an "Earthling" you have an obligation to think about this problem. Don't add to the population until you're ready to provide for your family.

Some governments try to control their population size by:

1. limiting the number of immigrants
2. encouraging emigration to other countries
3. decreasing birth rate (i.e. the "pill", sterilization)

Some countries, such as Sweden, the United Kingdom, and West Germany, for example, are near **zero population growth**. Zero population growth refers to the death rate and the birth rate overall being about equal. There is no net gain in population. The United States, Canada, Australia, and other countries are slowly decreasing in population growth. Other countries, like Mexico, are increasing rapidly.

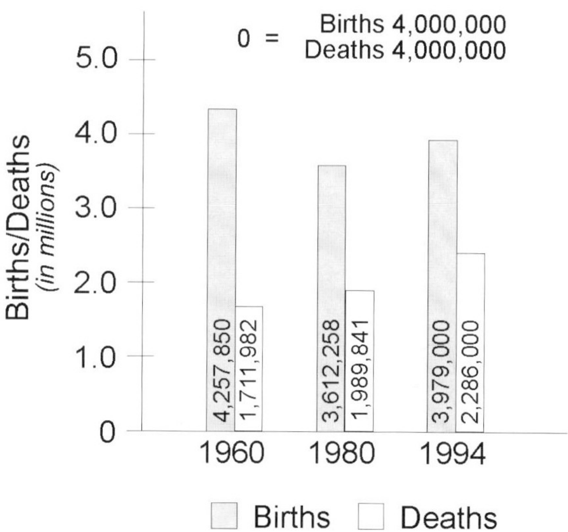

Figure 5.37: POPULATION GROWTH. Look at the three groups of birth and death-rate figures. As the birth rate and death rate get close to each other in number, zero population growth is inevitable.

SUMMARY

We see from the information above that not all countries are at the same point in population growth. Today the world has a better "feel" for the implications of a rapid growth rate. The depletion of the food supply, natural resources and space are inevitable. Our carrying capacity is limited. Some of nature's laws have been by-passed by humans, but sooner or later, the environment will force our hand on this issue, probably by diseases, natural disasters, or depletion of natural resources.

⬊ DID YOU KNOW?

The longevity of humans in Siberia is so much greater than other countries. Why would this be?

✔ THINGS TO KNOW AND DO:

Define:

* doubling rate ____

____

* zero population growth ____

____

*survival of the fittest ____

____

Explain the reasons for the shortened time period pertaining to the doubling rate of the human population. (Write a well-organized paragraph with at least three reasons.)

____

____

____

____

____

____

____

____

✎ IN SEARCH OF...

Using outside resources find the current population of:

The United States ______________________________

India ______________________________

Japan ______________________________

Switzerland ______________________________

Australia ______________________________

6

CARBON

OXYGEN

NITROGEN

STRONTIUM

ENDANGERED

RESOURCES

VARIETY

AREA

TECHNIQUES

IRON

ORE

NICKEL

VI. **CONSERVATION**

    A.   Definition

    B.   Groups of Natural Resources

        1.   Inexhaustible
        2.   Renewable
        3.   Nonrenewable
        4.   Recyclable

    C.   Types of Conservation

        1.   Soil Conservation
            a.   The Concern
            b.   Soil Conservation Methods

        2.   Wildlife Conservation
            a.   The Concern
            b.   Wildlife Conservation Methods

        3.   Water Conservation
            a.   The Concern
            b.   Water Conservation Methods

        4.   Forest Conservation
            a.   The Concern
            b.   Forest Conservation Methods

        5.   Mineral Conservation
            a.   The Concern
            b.   Mineral Conservation Methods

    D.   Summary

1.   Do not eat heavy foods before studying.
2.   Take notes, highlight or underline while studying.
3.   Try to study at the same time each day.

Things to Look for in This Chapter:

- Definitions for **all** highlighted words.
- The four types of natural resources and examples of each.
- The five areas that conservationists are studying.
- How mankind has helped and hindered wildlife, water, soil, forests, and minerals.
- The three levels of extinction.
- The different types of wells, soil, water, and purification methods.
- The different methods of harvesting forests.
- Mining techniques.

DEFINITION

When we hear the term **conservation**, we think of the root word "conserve." To conserve something means we use it carefully, wisely, and not wastefully. Conservation includes the wise use, the management, and protection of natural resources. It includes both abiotic and biotic resources such as sunlight, water, minerals, fossil fuels, plants, and animals.

Conservation does not mean that these natural resources are stock piled or hoarded but are used wisely to ensure their availability for future use.

Conservationists work in a wide range of activities and studies. Their main focus is to always provide the maximum use out of the minimum amount of resources. Included in the conservationists' concerns is the managing of forest, farmland, wild life and wilderness areas. Conservationists also search for ways to safely develop and use **natural resources** (minerals, air, soil, water, forests) and ways to meet the world's ever-increasing energy needs. Air pollution, water pollution, land pollution, and waste disposal also fall under the conservationists' care.

Natural resources can be broken down into four (4) main groups:

1. inexhaustible resources
2. renewable resources
3. nonrenewable resources
4. recyclable resources

Sunlight is one **inexhaustible resource** because it cannot be used up. Water is also in this group because the amount of water on earth remains constant. However, the supply and the cleanliness of fresh water may vary from place to place. Some minerals and salt exist in very large amounts on the Earth and are not likely going to be used up. Even with inexhaustible resources, our minds should be set on conserving.

Renewable resources, such as plants and animals, are organisms that can be used and replaced. Through the reproductive process these organisms are replaced by their offspring. Renewable resources are not stored for future use but harvested as needed. Trees, for example, if not cut down at the right time, may rot and become useless for the lumber indus-

Figure 6.1: CONSERVATION. In the top photo, farms are laid out to provide food, cover, and shelter for wildlife and to set aside natural land barriers. Between fields, unplowed lands containing hedges, bushes, trees, and grasses provide refuge for wildlife. The bottom photo shows a refuge, land set aside where water fowl and other wildlife can nest in winter.

Figure 6.2A: RESOURCES. The sun is one of our inexhaustible resources, as is water. This form of energy is reliable, environmentally friendly, and a domestic source of electricity. (left photo: U.S. Soil Conservation Service; right photo: U.S. Department of Energy).

Figure 6.2B: RENEWABLE RESOURCE. The planting and harvesting of switchgrass (biomass) are dedicated for fueling power plants. In days to come, biomass power plants will be located near large frams. The biomass (switchgrass) can be reduced to fine pieces for combustion in gas turbines to generate electricity. Gas turbines have a higher efficiency rate than steam turbines.

try, so they are harvested as needed. Renewable resources interact with one another since they are living organisms. The members of various food chains and food webs, as well as the abiotic factors that influence them, are all part of renewable resources. Soil, plants, and animals are influenced when trees are cut down. Animals may use the tree for shelter or food; plants on the ground once shaded by the tree now have abundant sunlight and even the soil quality is influenced by the tree whose root system has lowered the rate of soil erosion. The interrelationships are many if one takes a closer look. For example, let's think about sand dunes and the grasses growing on them. What would happen if the grasses were removed? Name some natural resources and interrelationships that might be affected.

____

____

____

____

Some natural resources are not renewable such as coal, iron, and petroleum. **Nonrenew-**

Figure 6.3: NON-RENEWABLE RESOURCE. Petroleum (gasoline), iron, and coal are not easily replaced. It takes many thousands of years to replace them, so we consider these resources non-renewable.

able resources, the third group, can not be replaced. They take many thousands or even millions of years to form. Today, these resources are being used faster than they are formed. These types of resources can be stored or left in the ground for use by future generations. Are there any other nonrenewable resources that you can list?

The fourth group of natural resources are **recyclable**. Paper, aluminum and glass are a few of the recyclable resources. For years schools and churches have had newspaper drives; now they have expanded their collecting to include aluminum cans. We would have been far better off in terms of reusing our natural resources if we had taken a tip from church newspaper drives.

Along our streets we find little storage bins for all types of recyclable materials. (See Figure 6.4) Motor oil, paper, glass, steel cans, aluminum cans, and plastic containers are some of the more popular recyclable materials that citizens of the United States recycle. Cities have initiated recycling programs and in most cases they have been met with enthusiasm by citizens.

The most important part of recyclable materials is the fact they may be reused over again. This reuse reduces the stress we place on the environment for needed natural resources.

Conservation practices continue to grow worldwide. Programs locally and nationwide are making the daily news. As public awareness grows in conservation issues, the future generations of the world will be provided for and the quality of life maintained.

One of America's first conservationists was John Muir (1834-1914.) He was a naturalist and explorer who traveled throughout the West. He established national parks and refuges for the preservation of American forests and wildlife.

TYPES OF CONSERVATION

The study of conservation may be divided into several groups:

1. soil
2. wild life
3. water
4. forest
5. minerals

Figure 6.4: RECYCLABLE RESOURCES. Paper, glass, and aluminum are a few of our recyclable resources. Recycle bins such as these line our city streets, providing convenient drop-off locations.

Soil Conservation

Soil on the Earth was and is made from rock over millions of years. Changing rock by wind, rain, glaciers, melting snow and even some plants into soil particles is called **weathering.** Weathering is both a chemical and physical process. The action of all these factors wears down rock into soil particles. This process is very slow, taking between 500 to 1000 years to produce 2.5 cm or about an inch of topsoil.

Topsoil is the upper level of soil containing organic matter (**humus**), rock particles, and many tiny organisms. The bacteria in the soil breaks down the humus into nutrients needed by the plants. It is estimated that about 60 million bacteria live in a single particle of topsoil. Some of the other organisms that make topsoil their home are insects, earthworms, mold, and yeast.

The layer of soil beneath the topsoil is subsoil. **Subsoil** is mainly rock particles but does contain tiny amounts of organic matter. This thin layer of topsoil that covers the Earth's surface is delicate. The soil forming process, as mentioned, takes many thousands of years. Man in some cases has destroyed topsoil in some areas in a very short time. For example, the rain forest is being destroyed at an enormous rate. Trees are cut down, the land cleared and plowed. The land is then farmed for a few seasons and left. More forests are cut and the cycle continues. What is left is a soil depleted of nutrients, and an area stripped of its natural

Figure 6.5: DUST STORM. The erosion of land stripped of vegetation results in the removal of nutrient-rich topsoil. (National Park Service)

flora and fauna. It takes a good many years to replace what was lost because the soil that is left is gradually worn away by erosion.

Erosion, the gradual wearing away of soil by rain and wind, usually occurs very slowly. When the land is stripped of vegetation, due to construction, mining, or farming, the soil is left bare and the rate of erosion increases. With

Figure 6.6: BEACH EROSION. Wind, water, and storms have eaten away protective sand dunes leaving a dangerous predicament for this homeowner. Over development of our beachfront property is a risky business. (David W. Arnold, Florida Department of Environmental Protection)

Figure 6.7: DUST BOWL. In the 1930's, Colorado, Kansas, New Mexico, Oklahoma, and Texas farmers, to capitalize on high wheat prices, did not allow the land to lie fallow (rest). This, along with the drought of 1931, caused the topsoil to erode and to be swept away by wind, darkening the skies for thousands of miles. (Biosphere 2000, p. 322)

plants removed, the soil receives the full direct force of the rain drops and wind. The soil once held in place by the root system of the plants is left at the mercy of running water and wind. The water running off the land, **runoff**, carries soil particles with it. When vegetation is in place, the plants absorb some water, help hold the soil particles in place, and therefore less soil is washed away.

Much of the eroded soil is deposited in lakes, streams, and rivers. In the United States about 175 million acres of land have been severely damaged by soil erosion. Most of the soil erosion took place in farmland. In the 1930's, the Great Plains region of the United States had a large amount of topsoil erosion. During these years very little rain fell on the plowed fields making the soil very dry and in prime condition for erosion. The grass that once covered the land was no longer there. Strong winds and dust storms blew and carried topsoil to the Atlantic coast and even out beyond the Gulf of Mexico. The land, with little or no soil left, was of no use to the farmers. Farms were abandoned and the land left to recover slowly.

In a previous chapter we discussed the lengthy amount of time needed for the formation of soil through succession. Today farmers reduce the rate of soil erosion by planting trees and leaving patches of natural vegetation between plowed fields. The trees and grass root systems help to keep soil in place by acting as wind breaks and reducing the runoff of rain water. As a result, the soil remains in- tact and productive.

The required nutrients found in the soil's humus are taken in by plant root systems as is water. Since plants are producers, they are very important to each of us. Soil must be kept in good condition for maximum plant growth. We will now look into the conservation methods that are practiced today, especially by farmers.

Soil Conservation Methods

There are various soil conservation methods used in today's world. These methods strive to have the single most important element in common: to conserve soil while using it. In doing do, we are acting wisely as good stewards, caring for the land given to us.

Soil conservation methods described adhere to the following basic conservation principles:

1. Keep soil covered, thus preventing topsoil loss.
2. Reduce the rate at which rain water runs off the land (run-off), thus preventing large amounts of topsoil and nutrients from being carried by water into ponds, lakes, streams, and oceans.
3. Increase the amount of rain water absorbed by the soil. The root systems of

plants absorb rain water in large amounts, reducing runoff.

4. Add mineral and organic matter to soil. Different plants, like legumes, add nutrients to soil. Fertilizers, like manure, increase soil fertility. Soil litter also adds to the soil's organic content.

These four soil conservation principles when practiced are effective in reducing the erosion of soil. The following are some of the most popular used soil conservation methods:

1. **Contour plowing**—This method is practiced on sloping land. Land is plowed back and forth across the slope rather than up and down. By plowing this way, ridges are formed and this slows down the flow of rain water.

2. **Strip cropping**—Strips of row crops are alternated on slopes. Close growing plants such as grass or clover are planted in strips between strips of corn, wheat or other grain crops. The grasses hold the soil in place better than do other crops.

3. **Cover crops**—After the main crop is harvested in the fall, farmers plant oats or grain in order to keep the soil covered and prevent soil loss.

Figure 6.8: NEED FOR SOIL CONSERVATION. Trees need enough soil for their roots to anchor themselves. Due to poor conservation methods and to land over development, much of the soil has been lost to erosion.

4. **Terracing**—Wide, flat rows called terraces are built along hillsides. Looking at a terraced hillside you would see a resemblance of a huge staircase. The terrace holds rain water, preventing it from washing soil away down a hillside and the formation of gullies.

5. **Crop rotation**—Soil fertility declines over time partly due to the replanting of the same crop year after year. Nitrogen

Figure 6.9: STRIP CROPPING. Close-growing plants such as grass and clover are planted in strips between corn, wheat, and other main grain crops. The grasses hold soil in place better than the grain crops, thus reducing erosion. (U.S. Soil Conservation Service)

Figure 6.10: SHELTER BELTS. Trees, or large, thick shrubs are planted in wide rows to break down the wind speed, reducing soil erosion. (U.S. Soil Conservation Service)

in the soil is taken in and used by plants such as corn, wheat and other grain crops. When legumes such as soy beans or alfalfa are planted, they restore nitrogen to the soil. Farmers rotate crops planted in a field; one year corn, the next, alfalfa. Rotation of crops keeps the soil fertility level high.

6. **Conservation tillage**—This method deals with the number of times a field is plowed each year. Usually a field is tilled three or more times a year. In **zero tillage** farmers leave the remains from the field crops as a covering for the soil. In the next planting the seedbed is prepared leaving the space between the crop rows with that residue still as a soil covering. Soil erosion is reduced as is tractor fuel usage.

7. **Hay and pasture land**—Farmers plant grasses and legumes for hay and pasture reducing soil erosion and adding nitrogen to the soil.

8. **Shelter belts**—Trees or large thick shrubs are planted in wide rows to break or slow down the wind speed. Lower wind speeds reduce soil erosion.

9. **Reforestation**—In areas where land is steep, stands of trees are planted that reduce the erosion of soil. The root system holds the soil particles intact; the tree itself cuts down wind speed.

10. **Dams**—Man builds dams and reservoirs to hold runoff water in times of heavy rainfall, preventing flooding and reducing soil erosion.

Farms may use one or more of these conservation methods according to the lay of the land, climate and soil type. Soil conservation helps to conserve our soil and its fertility and ensure adequate crop production for future generations.

Figure 6.11: DAMS. Dams hold back run-off water in times of heavy rainfall, helping to prevent flooding and soil erosion. (U.S. Soil Conservation Service)

↘ DID YOU KNOW?

The United States' largest capacity hydro plant (dam) is the Grand Coulee Dam in Washington. The highest dam in the U.S. is Oroville Dam in California.

✔ THINGS TO KNOW AND DO:

Define:

*conservation ____

____

____

*natural resources ____

____

____

*weathering ____

____

____

*topsoil ____

____

____

*subsoil ____

____

____

*erosion ____

____

____

*runoff ____

____

____

List 4 main groups of natural resources:

1. _______________________   3. _______________________

2. _______________________   4. _______________________

*Give an example of each of the above groups.

1. _______________________   3. _______________________

2. _______________________   4. _______________________

Name 5 groups being studied by conservationists.

1.____

2.____

3.____

4.____

5.____

➜ MODIFIED TRUE - FALSE: correct underlined word if necessary.

________ 1. Conservationists focus on providing the maximum use out of the <u>maximum</u> amount of resources. ________________

________ 2. Sunlight is an <u>exhaustible resource</u>. ________________

________ 3. Renewable resources are replaced through <u>reproduction</u>. ________________

________ 4. Renewable resources <u>do not</u> interact with one another. ________________

________ 5. Coal, iron, and petroleum are <u>inexhaustible</u> resources. ________________

________ 6. Glass is a <u>recyclable</u> resource. ________________

________ 7. <u>John Lennon</u> was one of America's first conservationists. ________________

________ 8. <u>Topsoil</u> is the home of many insects, earthworms, and bacteria. ________________

________ 9. About one million bacteria can be found in a single particle of topsoil. ________________

________ 10. The layer of soil beneath topsoil is called <u>subsoil</u>. ________________

✎ IN SEARCH OF...

Using the four basic conservation principles, describe in detail crop rotation and conservation tillage.

____

____

____

____

Who helped establish national parks and refuges for wildlife?

____

____

____

Wildlife Conservation

Wildlife includes all untamed animals and uncultivated plant life. Wildlife adds much beauty and enjoyment to our lives. At times we are amazed at the diversity of life that surrounds us. We recall sitting, glued to our chairs, while watching a documentary film on some form of wildlife. We became so involved in what we saw that we began to notice the complicated patterns some wildlife display in protecting their territory, shelter, or nesting sites. We even observed mating rituals that sometimes seemed strange and very extreme.

In these documentaries we saw plants and animals in the wild, living in their natural environment, exhibiting nature's balance. We learned that shelters, nesting sites, food webs, mating, and predator-prey relationships all function well if the balance remains undisturbed by man.

We see on the news every day that natural catastrophes, as well as man's influences, interrupt this delicate balance to some degree. It must also be noted that not all interruptions are negative but may have positive effects on wildlife. The damage resulting from severe storms, floods, or fires may create situations that benefit certain organisms.

Forest fires clear out a portion of the understory of an area. This allows growth of other organisms to sprout forth. Growth of new plant types that may not have had a chance to develop due to a slower growth rate now has the chance. Previously needed nutritional requirements or a specific amount of sunlight is now met in this situation. Pine cones of certain evergreen trees will only open and release their seeds under forest fire conditions. We, as co-inhabitors, must look at the various situations that arise in nature and try not to intentionally upset the balance, but let nature run its course.

Besides its beauty and enjoyment, wildlife is also an important part of our ongoing scientific research. From wildlife research man has found many uses for the materials that can be extracted from plants. **Quinine**, a substance derived from the bark of cinchona tree, is used to treat malaria. **Acetylsalicylic acid** (aspirin) made from salicylic acid from willow tree bark is used to reduce fever and pain. **Digitalis** is a heart stimulant obtained from the leaves of the

Figure 6.12: NATURAL CATASTROPHE. In a recent ice storm in Maryland (left photo), many miles of flora and fauna were destroyed, including wildlife shelters, nesting sites, mating rate, and predator-prey relationships. (L. Anderson) Right photo is a fernery in Pierson, Florida, where a cold snap threatened the plants. The ice was created on purpose by a sprinkler system in order to shield the plants from below freezing temperatures. The ice acts as an insulator. (middle photo: D. McBride; right photo: S. Nepton)

purple foxglove. Can you list several other ways wildlife has aided mankind?

____

____

____

Man, through carelessness, has contributed to the extinction of many plants and animals. In the past, uncontrolled hunting was a major cause of the near extinction of some organisms and to the extinction of others. Today we have laws that regulate the type and amount of wildlife that may be hunted. Hunting seasons are for a specific time period, and in some cases only one sex may be taken. Special permits are issued to hunt animals which ordinarily are not hunted, such as the alligator.

Besides over-hunting, habitat destruction by construction, rerouting of waterways, and destructive transportation methods have lowered certain populations of wildlife. Some wildlife now have less than adequate amounts of nesting sites, breeding areas, shelters and food because their habitats were plowed under by man's progress.

Water has been dammed, rivers have been rerouted, land has been plowed under, forests have been cleared, and swamps have been

About 25% of all pharmaceuticals used in the United States have ingredients from plants. Animals as well as plants provide useful medicinal products for man such as hormones, estrogens and thyroid extracts.

drained to make room for man. These activities, undertaken for man's growing population, as well as greed, have diminished, harmed and destroyed necessary habitats of many wildlife. As the number in a species has been decreased, and even become nonexistent, our variety in both plant and animal life forms has also decreased. We no longer have the variety of species that we once had. Once an organism becomes extinct, it is gone forever.

The destruction of rain forests and their wide variety of native wildlife species is a hot topic today. We are becoming aware of the alarming rate the forest is being destroyed by the "slash and burn" method; this clearing of land for a few, short, crop-growing seasons and then moving on is the plan at hand. We are well aware of the time it takes for nature to heal and replace what has been lost at man's

Figure 6.13: CHARRED PALM TREE. Fires, clearing underbrush and thinning out of shrubery, provide beneficial aspects for wildlife—providing more sunlight, new growth, and more room.

hands. About 50,000,000 acres a year of rain forests are being destroyed. At this rate it is feared that hundreds of plant and animal species will be eliminated.

As man moved into other wildlife habitats, pollution from industries, sewage, and pesticide contaminants built up within wildlife tissues. As time passed, various populations of wildlife were reduced due to the contamination.

Besides pollution, wildlife, such as snakes and alligators that play key roles in nature's balance, are being killed by man simply because man fears them. These predators control the population numbers of other wildlife; therefore, they do their part in the balance of nature, and we humans are upsetting this balance. For example, in September, 1995, in Melbourne, Florida, an alligator was noticed swimming in the inlet. The proper government officials were called, retrieved the alligator and destroyed it. How would a conservationist have handled this situation?

____

____

____

Wildlife conservationists are trying to correct the errors of the past to prevent the errors from recurring. Today the conservationists' goal of ensuring the survival of wildlife reaches out in many different ways and in many parts of our world. National publications are printed yearly giving free information to

Figure 6.14: LAST PASSENGER PIGEON. Contrary to previous belief, an over-abundance of a species does not mean it cannot become extinct. The last passenger pigeon died in 1914, despite the fact that it once had a tremendous population. (U.S. Forest Service)

the public about conservation efforts in different areas of the country. These publications may explain how an animal or plant became extinct or endangered. Details are given to explain the events which led to an organism's

Figure 6.15: ALLIGATORS. Alligators play key roles in nature's balance, but are being killed simply because man fears them. Can you think of other organisms that are killed because man fears them?

Figure 6.16: GAME LAWS. Today hunting and fishing are regulated by the U.S. Fish and Wildlife Service. Hunters may take a specific number of wildlife during the yearly set season. Regulated hunting benefits both man and wildlife.

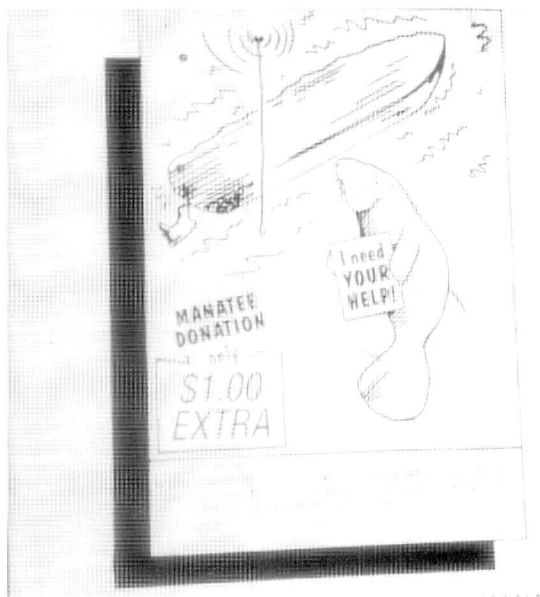

Figure 6.17. MANATEE DONATION. Money is collected in various locations throughout the state of Florida in support of manatee protection programs. Reprinted from the West Indian Manatee in Florida, copyright 1989, Florida Power and Light Company. By permission of the Florida Department of Environmental Protection.

extinction or to its being placed on the endangered species list. Also, these publications give suggestions that we, as good stewards, may use in order to correct the errors of the past.

Another way the conservationist reaches out is through **reforestation** programs that aid wildlife conservation. Planting shrubs and trees provides cover for many forms of wildlife. The fruit and nuts that develop from the planted trees provide additional food. Wildlife, through reforestation, become adequately protected as well.

As previously mentioned, fish and game laws aid wildlife conservation by protecting the numbers of animals that may be taken. Restrictions are set up that allow hunting and fishing to take place at **specified times** of the year, for a **specified organism**, and for a **specified number**. Stiff penalties and sentencing result for those who refuse to abide by these rules. The job of state game wardens and other paid officials help keep the poaching down. **Poaching** is, however, an ongoing problem. Taking of wildlife that are endangered or not in season reduces significantly the effectiveness of any conservation program.

One such program is directed toward our waterways as they became a major recreational spot. Speed limits are now posted where endangered or protected wildlife exist. The manatee population has decreased in Florida

Figure 6.18: SAVE THE MANATEE. The Florida Department of Natural Resources receives funds when car owners purchase this type of automobile plate. These funds go to research and conservation activities at state and federal levels. Reprinted from the West Indian Manatee in Florida, copyright 1989, Florida Power and Light Company. By permission of the Florida Department of Environmental Protection.

waters partly due to motor boat propellers. These slow-moving creatures often fall victim to speeding motor boats. A large number of existing manatees show propeller scars on their bodies. These are the fortunate ones, the ones that survived their brief encounter with man. A careful regard on our part in the direction of wildlife will ensure the manatee's survival.

Wildlife refuges, fish hatcheries and wildlife research are maintained by the **U.S. Fish and Wildlife Service**. As well as animal conservation, the U.S. Fish and Wildlife Service strives to conserve plants, soil, and water. These three resources benefit and add to the survival of wildlife.

Boating Speed Zones

Safe Operation Zone - a sign indicating that you may resume safe boating speed; visible as you leave a protected area.

Caution Zone - a zone frequently inhabited by manatees, requiring caution on the part of boaters to avoid disturbing or injuring the animals.

Slow Speed Zone - a no-wake or minimum-wake zone where boats must not be on a plane and must be level in the water; generally these signs are posted on the fringe of protected areas to warn you that you are approaching an area frequented by manatees; in some areas the channel is exempt.

Idle Speed Zone - a zone in which boats are not permitted to go any faster than necessary to be steered; generally these signs appear near the center of a protected manatee zone.

No Entry Zone - a protected zone that prohibits boating, swimming and diving for the protection of manatees.

Figure 6.19: MANATEE/BOATING SPEED ZONES. To alert the boater and protect the manatee, the law provides a number of cautionary and regulatory speed zones. At left are some illustrations and a brief explanation of the various signs. Illustration by Laura Sartucci Wiegert. Reprinted from the West Indian Manatee in Florida, copyright 1989, Florida Power and Light Company. By permission of the Florida Department of Environmental Protection.

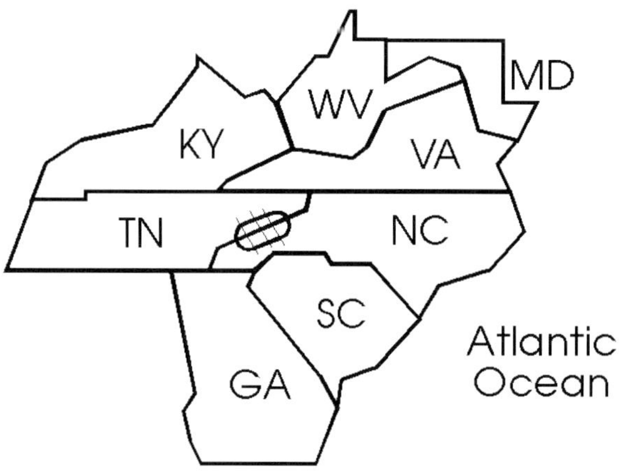

Figure 6.20: GREAT SMOKEY MOUTAIN NA-
TIONAL PARK. The shaded area on the Tennes-
see/North Carolina border is the location of the
800 square mile park.

We have national forests and parks which
provide a protected environment for wildlife.
The Great Smokey Mountains National Park is
an example of a protected environment. Covering over 800 square miles, the park protects
endangered red wolves, once scarce white-
tailed deer, and black bear. Biscayne National
Park, covering 181,500 acres, is 95% underwater. The park houses an endangered coral reef,
manatees, crocodiles, sea turtles, and red mangroves that were heavily damaged by Hurricane Andrew in 1993.

People have a place to go and enjoy nature
in the many parks and forests in the United
States. As you pass through life, visit these
areas, enjoy what you see, what you hear, and
what you smell, but always remember **take**
only pictures and **leave** only footprints.

Our country's awareness and development
of conservation programs have helped save
some species from extinction. Many rare
plants and animals, including the American
bison, for example, have been saved by such
programs. However, several hundred species
of plants and animals combined are in danger
of extinction. The blue whale, the Bengal tiger,
the California condor and the St. Helena redwood, to name a few, are endangered species.
(See partial list, Figure 6.24)

Figure 6.21: BISCAYNE NATIONAL PARK.
Covering 181,500 acres, the park is home to endangered coral reef, manatees, crocodiles, sea
turtles, and red mangroves that have extensive
prop root systems.

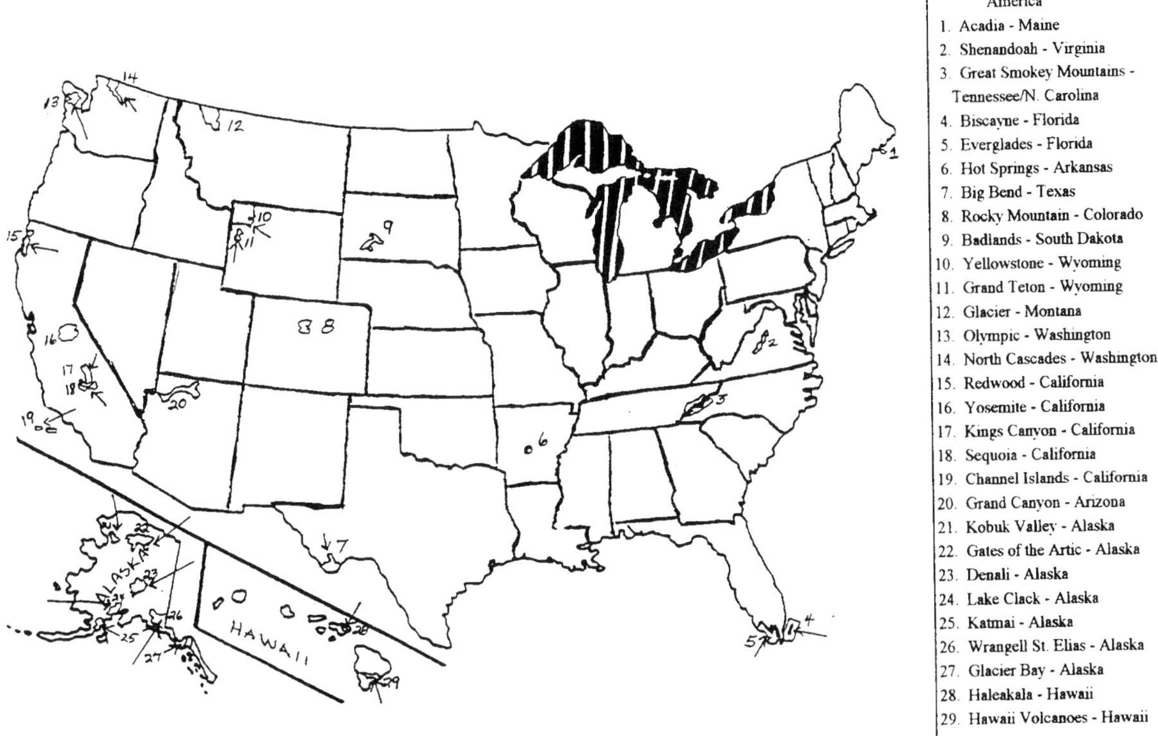

Figure 6.22: A FEW OF OUR NATIONAL PARKS. The U.S. government has set aside many parks and forests where people can go and enjoy nature.

Figure 6.23: CAMPING. The above photos show a primitive method and a more luxurious method of camping. These campers are drilling for water near their camp site.

Wildlife may be classified into three groups if they face a possible extinction.

1. **endangered**—The wildlife are seriously facing extinction.
2. **threatened**—The wildlife may be abundant in some areas but they are still facing serious dangers. (example—over hunted or fished or collected)
3. **rare**—The wildlife has small populations but numbers are not decreasing. They exist in protected areas, and their survival is not threatened.

So, depending upon location and numbers, specific species of wildlife are protected by conservation programs. The value of wildlife conservation is endless. The diversity and beauty of each specie of plant and animal heightens our concern for the protection of wildlife. Camping, hiking, boating, and other outdoor activities become more enjoyable when we are surrounded by wildlife.

Besides their esthetic value, wildlife has an economic value important to our economy. Wildlife provide income for our nation from revenues collected at zoos and refuges. Important plant products, food sources, and furs and skins provided from wildlife are also valuable resources for man.

Various life processes are also studied by observing and researching wildlife. We have gained much medical knowledge through research, and we understand some of the effects pollution has on living things. We are blessed in the variety of wildlife that is upon the Earth, and we must strive to maintain its existence. The need for all of us to be conservationists is imperative.

Partial List of Threatened and Endangered Species in the U.S.A.

Mammals

1. gray bat
2. American black bear
3. grizzly bear
4. Pt. Arena Mt. beaver
5. woodland caribou
6. eastern cougar
7. northern swift fox
8. mountain lion
9. West Indian manatee
10. ocelot
11. Florida panther
12. lower Keys rabbit
13. sea lion
14. California squirrel
15. gray whale
16. red wolf

Birds

1. bobwhite
2. condor
3. whooping crane
4. Hawaiian hawk
5. Everglade kite
6. owl
7. warbler
8. woodpecker

Reptiles

1. American alligator
2. American crocodile
3. Plymouth turtle

Figure 6.24: THREATENED AND ENDANGERED SPECIES. This list fluctuates as wildlife becomes endangered or extinct. Wildlife may be removed from this list when populations increase and become stable. (Fish and Wildlife Service, 1994)

List 10 wildlife products of economic value.

1. _______________________
2. _______________________
3. _______________________
4. _______________________
5. _______________________
6. _______________________
7. _______________________
8. _______________________
9. _______________________
10. _______________________

Note:

According to the National Wildlife Federation there are 500 wildlife refuges in the U.S. (1995)

↘ DID YOU KNOW?

The continent of Africa is home to millions of wildlife, most of which are protected on wildlife reserves, in conservation areas, and at national parks. More than 2% of African land protects its wildlife populations. Some noted parks are the Serengeti National Park and Mt. Kenya National Park.

✔ THINGS TO KNOW AND DO:

Define:

* wildlife ____

____

____

* "slash and burn" _______________________________________

____

____

*reforestation _______________________________________

____

____

*U.S. Fish and Wildlife Service _______________________________

____

____

✍ FILL IN THE BLANKS:

1. ________________ ________________, as well as man's influences, interrupt nature's balance to some degree.

2. Once an organism becomes ________________, it is gone forever.

3. ________________and ________________are being killed simply because people are afraid of them.

4. An ________________specie is wildlife that is seriously facing extinction.

5. A ________________specie is one that may be abundant in some areas, but still facing serious dangers.

6. ________________wildlife has small populations but numbers are not decreasing. They exist in protected areas and survival is not threatened.

✎ IN SEARCH OF...

Write a short essay explaining man's positive as well as negative effects on the wildlife on Planet Earth. Make sure you use at least two positive effects and two negative effects.

____

____

____

____

____

____

____

____

____

____

____

____

____

____

____

____

____

____

Water Conservation

Water is required for all living organisms. Plants and animals use water daily in maintaining their life processes. The human population needs water to drink, bathe, cook and clean. Each person consumes about 760 liters of water per year, over 2 quarts a day. A shortage presents physiological problems that may lead to death if not corrected. **Dehydration** is the term used to describe the loss of water from the body. Usually in dehydration a loss of salt also occurs. Symptoms of dehydration may be dryness of mucous membranes of the nose, mouth and throat, and a reduced ability to sweat and urinate. In severe dehydration a rapid heartbeat, low blood pressure, shock and death may occur. The importance of potable (drinkable) water cannot be minimized.

In agricultural areas, farmers need water for their crops and for irrigation of dry land. Many of our industries and thriving cities have developed along the waterways. Water is also used to generate electricity. Niagara Falls, in New York State and on the borders of Canada, is a thriving waterfall that is used by both countries to produce electricity. Niagara Falls is located on the Niagara River and naturally consists of two waterfalls, Horseshoe Falls and American Falls. Here hydroelectric plants divert water through tunnels from the Niagara

Figure 6.25: NIAGARA FALLS, NEW YORK. One of the natural wonders of the world, with enough water to generate electricity to light half of the East Coast, Niagara Falls borders Canada and New York State.

River. These power plants harness water-power traveling through the tunnels to generate the electricity for a wide area, in fact, most of the Eastern seaboard.

The demand for fresh, clean, potable water is ever-increasing. As the world's population increases, the demand for water also increases. Growth and expansion of industry and agriculture increase the demand for water as well. Water is also important for transportation of goods and people. The Mississippi River, the Chesapeake Bay, the Great Lakes, the Nile River, to name few, are great waterways that affect the immigration and settlement of people, the trade of goods and services, and the movement of these products from one location to another. Many recreational activities involve water as well, such as swimming, boating, skiing, fishing and the list goes on.

The Earth does have an abundance of water. However, the water is not all usable in the form in which it exists, and it is not evenly distributed throughout the Earth. There are places where rainfall is minimal or excessive. From the Midwest to the West coast of the United States, water is limited. People in these areas drill wells to obtain water.

There are different types of wells that tap into the underground water supply. A **deep well** is drilled deep into this underground water supply. Some of these wells yield thousands of liters of water a day for the surrounding farms, towns, and industries. A **shallow well** is drilled into the **water table**, water deposited close to the surface of the land. The water table depth varies in different parts of the world. It could be as deep as 9 meters in some areas or just at the surface in others. In a lake, pond, or swamp, the water table is at the surface and not underground. Shallow wells are used in many areas to provide water for individual home owners.

Ground water is water found in pores of rock and sediment in fully saturated areas. Ground water is used less than surface water but is a very important water source for farm and domestic uses; it provides water for about one-half of the population of the U.S. In the Middle East, ground water is the only source of water for irrigation. In places such as Florida, Japan, and England where conditions are

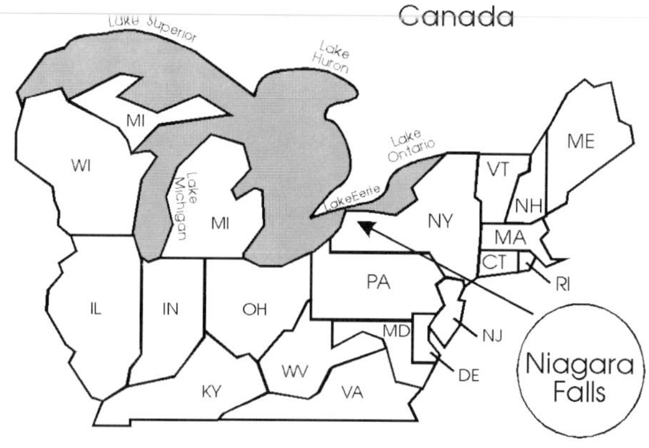

Figure 6.26: NIAGARA FALLS. As you can see, the Falls are located on the New York State and Canadian border, between Lake Erie and Lake Ontario.

humid, ground water is used, and preferred, to surface water. Ground water has less contaminants and is usually easy to obtain.

Water collected from lakes and rivers must be purified before it can be consumed by man. Large purification plants are used to remove unwanted material from the water. This unwanted material may be decaying plant and animal tissue, soil particles, and microscopic organisms. With the addition of chlorine at these plants, the water is cleaned and purified.

The building of **dams** provides needed water for human consumption in certain areas of the world; however it is sometimes at the expense of wildlife. One such occurrence happened in the late 1970's with the construction of the Tellico Dam on the Little Tennessee River. Conservationists felt that the building of the dam would have a harmful effect on the rare snail darter fish. The dam was built. The conservationists transferred the snail darters to the Hiwassee River in Tennessee which ultimately ensured their survival.

Dams also change the amount and flow of water traveling within a specific area. This change in a waterway may harm the fish that live there. Fish as well as most other living organisms require a specified set of livable conditions; however when these conditions are not met because of the major changes created

Figure 6.27: WATERWAY. Water is transported from a plentiful location to an arid one where it is needed. (U.S. Soil Conservation Service)

by the dams, survival of the organisms that inhabit the river is jeopardized. Organisms may not have the ability to adjust to the newly created conditions and therefore perish quickly.

Dams are not built in areas where the water contains an abundance of silt. When water containing high amounts of silt is dammed and goes into a reservoir basin that collects and stores rain water, the reservoir becomes quickly filled with silt, reducing the area for storing water.

Now let's look at the positive side of man's water conservation efforts. Dams and reservoirs prevent flooding during spring rains. Water conservation is also helped by the use of canals and aqueducts that transport water from where it is abundant to where it is scarce. One such place where a large aqueduct has been built and used is in California. There, aqueducts carry water from Northern California to Southern California, where it is needed.

Watershed management is another way of increasing water supply for man by reducing runoff of rainwater and melting snow. Plant life, grass, bushes, and trees cover the land and prevent water from running rapidly and carrying soil particles along. As a result of watershed management, more of the water is absorbed and filtered into the ground. The ground water now is replenished to a greater degree, and, as a bonus, soil is conserved at the

same time. The water flows underground back into lakes and rivers. When man removes plant cover for various reasons, water is not absorbed readily, soil is removed, and flooding may occur to a greater degree. Plants are an important part of the water cycle, helping to

Figure 6.28: WATERSHED. Top photo shows a natural watershed. (U.S. Soil Conservation Service)

absorb and direct our precious water and reducing soil erosion.

Some methods of water conservation are very costly and are considered not practical; therefore, they are only used to a small degree. **Desalination**, the removing of salt from seawater, is one such method. Scientists feel with the unlimited solar energy available, one day desalination will be commonplace. Conserving water is an important issue in today's world, not only for ourselves, but for future generations.

Can you think of ways you and your family can conserve water?

____

____

How about your town?

____

____

What positive results will we see if we become good conservers of water?

____

____

____

____

____

↘ DID YOU KNOW?

It takes 246,000 liters of water for the production process of one ton of steel (about the weight of a midsized truck).

✔ THINGS TO KNOW AND DO:

Define:

*dehydration ____

____

____

*potable water ____

____

____

*deep well ____

____

____

*hydroelectric plants ____

____

____

*aqueduct ____

____

____

*desalination ____

____

____

*reservoirs____

____

____

*snail darter fish ____

____

____

Name the famous falls that exist in northern New York.____

Why are they important?

____

____

List some ways you use water at home, at school, at leisure time, etc.

1. ____
2. ____
3. ____
4. ____
5. ____

✎ IN SEARCH OF...

Explain the difference between surface water and ground water.

____

____

What are purification plants used for? Where does the water come from?

____

____

____

What are some symptoms to look for when you suspect someone is dehydrated?

FOREST CONSERVATION

In the United States forests cover about 1/3 of the total land area. Forests serve as homes and cover for many forms of wildlife. Forests aid in conserving soil, preventing erosion and furnishing food for wildlife. Breathtakingly majestic forests provide recreational areas for camping, hiking, wildlife watching, and hunting. Forests are also important as watershed areas by absorbing large amounts of water and reducing rapid runoff. Forests supply many commercial products such as timber for construction and wood pulp for making paper. Resins, dyes, and turpentines from forests are used in various industries.

Stop and take a quick look around you right now while you are sitting and reading this book. What do you see that may have come from a forest? Make a list of items that are in some way linked with forests. Name 15 items.

Figure 6.29: CALIFORNIA REDWOOD FOREST. The majestic beauty of the giant redwoods leaves one speechless. (U.S. Forest Service)

1. ___________________________

2. ___________________________

3. ___________________________

4. ___________________________

5. ___________________________

6. ___________________________

7. ___________________________

8. ___________________________

9. ___________________________

10. ___________________________

11. ___________________________

12. ___________________________

13. ___________________________

14. ___________________________

15. ___________________________

There are many types of forests worldwide. **Hardwood forests** found in the northeast United States contain maple, hickory and oak trees. From these forest we retrieve many food and lumber products. See Figure 6.24. **Evergreen forests** are found in Canada, Russia, and Alaska, containing spruce, balsam fir, pine, cedar, hemlock and Douglas fir. The giant redwoods, ponderosa pine, Douglas fir, white pine, Engelmann spruce, and white fir are found in California, Washington, and Oregon. A great supply of softwood lumber comes from the Pacific Northwest. In Southeastern California grow great stands of yellow pine. In Southern California **chaparral forests** are found, containing broad-leafed evergreen shrubs and a variety of small trees. In South Atlantic and Gulf States loblolly and slash pines flourish.

Many of the forests in the United States are **national forests.** These forests are owned and operated by the federal government. The United States Forest Service manages the forests with several purposes in mind: to provide shelter for wildlife, to furnish timber, to conserve water, and to provide recreational areas for use by American citizens.

Forest conservation practices include forest-fire prevention, forest-fire fighting, controlled burning, improvement cutting, selective cutting, block cutting, and reforestation.

Many forest fires are caused by careless hunters, campers and hikers. Camp fires left unattended or not completely put out and ciga-

---

PARTIAL LIST OF FOREST PRODUCTS

1. WOOD PRODUCTS
   a. Lumber (softwood and hardwood)
      *baseball bats   *molding
      *boats   *railroad ties
      *cabinets   *musical instruments
      *doors   *construction
   b. Particle Board (wood shavings mixed with adhesive, pressed at high temp and pressured to form large sheets)
      *drawer sides
      *cabinet tops
      *furniture
   c. Round Timbers (trees stripped of branches and treated)
      *bridges
      *fence posts
      *pilings for docks
      *utility poles

2. FIBER PRODUCTS
   (wood fibers compressed)
   *hardboard (furniture, siding, paneling)
   *insulation board (acoustical tile)
   *paper (bags, books, cartons, tissue, etc.)
   *paperboard
3. CHEMICAL PRODUCTS
   *cellulose (adhesives, lacquers, and plastics)
   *lignin (inks, dyes, concrete)
4. FUEL PRODUCTS
   *charcoal   *pulverized fuel
   *fireplace logs   *sawdust
   *wood chips
5. OTHER PRODUCTS
   *bark (adhesives, cork, dyes)
   *sap (maple syrup)
   *fruit and seeds
   *gum (pine oil, rosin)
   *leaves (holly, household, perfumes)

Figure 6.30: SOME FOREST PRODUCTS. We use materials from the forest in almost every aspect of our lives. Check out the list above with the list you made.

rettes thrown down on the forest floor are excellent forest fire starters. Nature has its part in creating forest fires. Lightning and very dry conditions are responsible for many forest fires. On the other hand, about 1/4 of all forest fires are caused by **arsonists**, people who purposely set fires. It is illegal to set fires and stiff penalties are imposed for arsonists. Forest fires burn thousands of acres a year and kill wildlife who live there.

In some instances, however, forest fires can be beneficial. Clearing out dense undergrowth by **controlled burning** keeps the forest "clean" and also minimizes the effect of destructive forest fires. Fires also prevent succession in a forest from reaching a climax community. The pine trees in Southeastern United States forests are harvested for the lumber industry. Less valuable hardwood trees, if not destroyed by controlled burning, would replace the pines. The pines are not harmed in these fires.

Forests are also aided by **improvement cutting.** In this method, diseased, crooked, or aged trees are removed. Less valuable species are also removed allowing more room for the more commercially desirable ones to grow. In **selective cutting** only aged trees are removed. With improvement and selective cutting higher yields of quality trees result.

In **block cutting** trees are harvested in large block-like sections. Seeds from the surrounding trees land in this open area, and, in time, the land is reforested by nature's hand.

Forests are also renewed by man planting seedlings (small trees) in sections of forests where trees were harvested. Lumbering companies will reforest land after they have harvested to ensure trees for future generations.

Figure 6.31: IMPROVEMENT/SELECTIVE CUTTING. Only aged, crooked, or diseased trees are removed, leaving the younger, rapidly growing trees behind. (U.S. Soil Conservation Service)

The government grows trees in nurseries which are used in **reforestation** programs. Many local private organizations have taken it upon themselves to reforest the land. They ask for contributions from citizens and for volunteers to help their reforestation programs.

With the improved methods of forest conservation in the United States, the destruction of our forests is being slowed. Programs such as reforestation, controlled burning, and selective cutting ensure future generations of having forests and the needed forest products.

⬂ DID YOU KNOW?

The National Forest Service administers over 189,000,000 acres of land in the United States.

Careers in Conservation

Forest ranger—protection and care of state and federal forest.
Air quality engineer—design and installation of air pollution control equipment.
Ecologist—scientific study of the relationship between living things and the environment.
Soil conservationist—planning and use of soil conservation methods.
Water quality technician—test water for pollution.
Forestry technician—assist foresters in care and management of forest lands and their resources.
Park technician—help plan and maintain day to day operations of a park.
Soil scientist—collect soil samples and classify them.

Career Info

Forestry technician
U.S. Department of Agriculture, Forest Service
Washington, D.C. 20250

Park Technician
Colorado Mountain College
Glenwood Springs, CO 81601

Ranger Manager
Society for Range Management
2760 W. 5th Ave.
Denver, CO 80204

Soil Scientist
Soil Society of America
677 S. Segoe Rd.
Madison, WI 53711

Figure 6.32: INDUSTRIAL USES OF SOME MINERALS. The above chart lists some minerals and their uses (given in percentages). (Biosphere 2000, U.S. Bureau of Mines)

✔ THINGS TO KNOW AND DO:

Define:

*national forests ____

____

____

*arsonists ____

____

____

*reforestation ____

____

____

*U.S. Forest Service ____

____

____

List six (6) ways that forests are being conserved.

1. ____
2. ____
3. ____
4. ____
5. ____
6. ____

Name the four groups of forest products and list 5 items in each group.

A. ________________________________

    1. __________________________
    2. __________________________
    3. __________________________
    4. __________________________
    5. __________________________

B. ________________________________

    1. __________________________
    2. __________________________
    3. __________________________
    4. __________________________
    5. __________________________

C. ________________________________

    1. __________________________
    2. __________________________
    3. __________________________
    4. __________________________
    5. __________________________

D. ________________________________

    1. __________________________
    2. __________________________
    3. __________________________
    4. __________________________
    5. __________________________

✎ IN SEARCH OF...

Using the map provided, place the following where they may be found:

1. redwood forests
2. hardwood forests
3. maple, oak, and hickory forests
4. evergreen forests
5. yellow pine forests
6. Ocala National Forest
7. George Washington National Forest
8. White River National Forest

MINERAL CONSERVATION

Minerals are matter found in the Earth's crust and oceans. Minerals are used in the production of many products. Cars, computers, buildings, airplanes, stoves, kitchen utensils and other items are produced from mined minerals. Some mined minerals are iron, lead, copper, zinc, gold, silver and nickel. The use of these naturally occurring minerals is on the upswing. Nickel, aluminum, copper, lead, and zinc are minerals that are now in abundance, but at the rate at which we are using them, they may become depleted in future years.

Mineral deposits are spread out unevenly throughout the world. The desire for minerals such as gold has led some countries toward exploration and the conquering of new lands. In our own country, the pursuit of gold led to the gold rush of Alaska and California which helped populate those areas. Some minerals we need we must import from other countries. We import mica, strontium, chromium, cobalt, fluorine, manganese, nickel, platinum, tin and others. (See Figure 6.33) The environment suffers as we mine and refine minerals that we depend upon. Some of our beautiful land, waterways, and wildlife have been destroyed by careless mining techniques. Mining copper, for example, may leave large open pits across the surface of the area mined, while other mining completely strips the top layers of land. After the minerals are mined, the refining process may add pollutants to the water, soil, and air. The discharge from refineries, besides contaminating water, kills all forms of wildlife.

The mining technique used is usually based on where and how the mineral deposit is found. Minerals may be deposited deep in the ground or close to the Earth's surface, scattered throughout an area or in one large deposit. Minerals may be found in ocean waters, rivers, and lakes. Methods of surface mining include:

1. open pit mining
2. strip mining
3. quarrying
4. dredging
5. placer mining

Mineral	Industrial Uses
Aluminum	Packaging (39%), transportation (20%), building (14%), electrical (8%), consumer durables (8%), other (11%)
Beryllium	Nuclear power, aerospace (40%), electrical (36%), electronic components (17%), other (7%)
Chromium	Metallurgical (52%), chemical (33%), refractory (15%)
Cobalt	Superalloys (37%), magnetic materials (16%), driers (11%), catalysts (10%), cutting and mining bits (7%), other (19%)
Copper	Refined metal fabrication (80%), other (20%)
Industrial diamonds	Machinery (27%), stone and ceramic products (22%), abrasives (16%), construction (13%), mineral service (8%), transportation (6%), other (8%)
Gold	Jewelry and art (61%), industrial (29%), dental (9%), small bars (1%)
Lead	Batteries and gasoline addItives (75%), construction, paint, ammunition (2%), other (5%)
Molybdenum	Iron and steel production (75%); machinery, oil and gas industry, transportation, chemical, electrical (25%)
Nickel	Stainless and alloy steel production (45%), nonferrous alloys (30%), electroplating (15%), other (10%)
Platinum group metals	Automotive (33%), electrical (28%), chemical (15%), dental (9%), other (15%)
Silver	Photographic (39%), electrical (29%), silverware and jewelry (14%), alloys and solders (7%), other (11%)
Tin	Containers (25%), electrical (17%), construction (13%), transportation (14%), other (32%)
Tungsten	Metalworking and construction machinery (72%), transportation (11%), lighting (8%), electrical (5%), other (4%)
Zinc	Construction (40%), transportation (20%), machinery (12%), electrical and chemical (15%), other (13%)

Figure 6.33: IMPORTED MINERALS. The United States imports many minerals necessary for industrial, commercial, and personal use. Notice the percents and how heavily we rely on some countries. (Biosphere 2000, P. 387)

Open pit mining, as used in the mining of copper, recovers minerals that are close to the Earth's surface. The top layer of rock and other material is removed, exposing the mineral deposit. The mineral and rock deposit is broken up by the use of explosives. The mineral and rock debris is picked up and hauled away.

The second method is **strip mining**. Miners cut a furrow, placing the removed top layer in a ridge at the sides of the furrow. Machines and explosives break up and remove the ore, placing it onto trucks or railroad cars. Phosphate and coal are mined this way. It was strip mining that caused great destruction in the past. Vegetation on mountain sides was removed, leaving the land bare, causing severe soil erosion and mudslides.

The third way is **quarrying**, a process that removes large chunks of matter that are found close to the surface level. Mica, gypsum, limestone, gravel, sand, and large rocks are examples, to name a few, that are mined in a quarry. Material is hauled by train or trucks to refining areas.

Dredging, the fourth method, is used where sand and gravel layers are very thick. Ponds or lakes must be formed so a dredge, bargelike machine, can be floated. The dredge dips a

Figure 6.34: MINING TECHNIQUES. The photo shows a strip mining camp. (Ransome, F. L, 526, U.S. Geological Survey)

bucket into the water. The bucket digs up sand and gravel that contains minerals which then are deposited and washed. The mineral is collected, the sand and gravel is placed to the rear of the dredge. Coal is sometimes mined this way.

The last surface mining method is **placer mining**. This method removes heavy metals such as gold, tin, and platinum from sand and gravel deposits. There are various placer methods, but for the most part, water is used to wash the sand or gravel. The heavy metals sink to the bottom level of a gravel box where they are collected.

Now let's switch from surface mining methods to underground mining methods which are many but are similar. Underground methods are used when mineral deposits are deep below the Earth's surface. First a shaft (a vertical opening) is dug along with horizontal passageways called levels. The minerals mined in these ways are lead, limestone, salt, potash, uranium and coal, which is still mined this way in the Appalachian regions which include Kentucky, West Virginia, Ohio, Tennessee, Pennsylvania and Alabama.

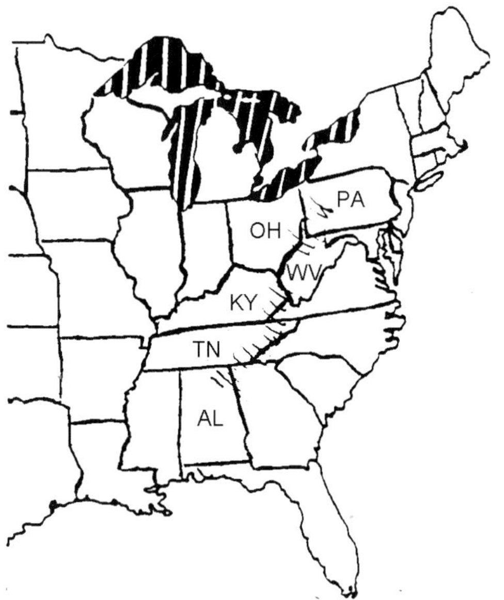

Figure 6.35: UNDERGROUND MINING STATES. In the Appalachian regions shown above, coal, uranium, and other minerals are mined below ground.

With all of the mining in the United States, conservationists are working hard to maintain and protect wildlife. Today mining companies must have a plan for restoring the mined land to its original condition before they are allowed to mine. This process that must occur after mining is completed is called **reclamation.** Conservation of minerals also includes reducing the amount of waste resulting from mining and refining methods. We are now more efficient than in the past. Currently minerals are used as substitutes for others in order to protect any one mineral from being depleted.

Obtaining minerals from recycled products is energy efficient because we now make new products from these previously used minerals. Aluminum cans are a popular recycled material as well as nickel, chromium, lead, copper, and zinc which are found in discarded autos, used machinery, and old appliances.

SUMMARY

Remember, reuse, reduce, conserve, and preserve. These are our common goals as good "stewards." The recycling bins you have in your neighborhood are for your use.

Figure 6.36: RECYCLING. All the above photos show recycling methods: Top—curbside pickup of aluminum cans; The second and third photos show old, rusted, farm equipment being refurbished. Both methods reduce the depletion rate of mined minerals.

↘ DID YOU KNOW?

Aldo Leopold, an American naturalist, influenced the American conservation movement in the early 1900's by writing the first wildlife management textbook, *Game Management*.

In 1973, the Endangered Species Act was passed which provided more protection for threatened and endangered species of wildlife.

✔ THINGS TO KNOW AND DO:

List seven (7) minerals that are mined.

1. _______________________   5. _______________________

2. _______________________   6. _______________________

3. _______________________   7. _______________________

4. _______________________

Define:

*open pit mining ____

____

____

*strip mining ____

____

____

*quarrying ____

____

____

*dredging ____

____

____

*placer mining ____

____

____

*reclamation ____

____

____

MATCHING:

_________ 1.　open pit mining　　　a. gold, tin

_________ 2.　strip mining　　　　　b. lead, uranium

_________ 3.　placer mining　　　　c. coal

_________ 4.　dredging　　　　　　d. copper

_________ 5.　quarrying　　　　　　e. mica, gypsum

_________ 6.　underground mining　f. phosphate and coal

7

POLLUTION

RECYCLING

VII. ENVIRONMENTAL POLLUTION AND RECYCLING

A. Pollution

1. Introduction

2. Air pollution

3. Water pollution

4. Soil pollution

5. Solid waste pollution

6. Other pollutants

B. Recycling

C. Summary

1. Remember the 3 R's:
    Read
    Record
    Recite
2. Have your own study space.
3. Try some practical application for each new idea you learn.
4. Set a goal daily for yourself and try to keep it.

Things to Look for in This Chapter:

- Definitions of **all** highlighted words.
- The five types of pollution.
- Causes of each type of pollution.
- Remedies for each type of pollution.
- Recyclable materials.
- Governments' role in recycling. (local and federal)

POLLUTION

Within our everyday lives we humans, as inhabitors of the Earth, pollute our environment by our daily activities. We contaminate our atmosphere with gases from our cars, buses, factories, and cigarettes; we pollute our waterways with dumping of trash and oil spills, as well as fertilizers and pesticides that are washed from the land; we pollute the land by throwing trash and by putting pesticides and fertilizers into the soil; we even cause noise pollution with the running of heavy machinery, loud stereos, and loud vehicles. In some way we all pollute.

Pollution is an ongoing serious problem. We need air, water and soil to survive. As we use resources to any degree, we pollute the environment. Air pollution can cause sickness and death, especially to those who have respiratory problems. Water pollution kills the organisms that live in ponds, lakes, and oceans. Soil pollution contaminates the land which reduces the amount of usable land for crop production. As pollution continues, the esthetic value of the Earth decreases.

We all are trying to do our part in reducing pollution; however, there is a "hitch" in reducing pollution; everything that benefits people pollutes the environment. Factory wastes, automobiles' and buses' exhaust, farm fertilizers and pesticides are all pollutants, but all these items benefit man in some way.

Figure 7.1: YUCK! POLLUTION! The graphic depicts various types of pollution. What could you add to this graphic? (Draw it in the space provided.) Circle the items **you** should eliminate from your environment.

In early times pollution was not much of a problem because of the low population and small amount of polluting items. The industrial revolution in the 1700-1800's led the way for pollution to take hold. The increase in goods produced also increased the rate of pollution. By the mid 1900's, pollution had shown its effect in air, water and land. People's awareness of pollution increased, and we began working on our pollution problem.

Many forms of environmental pollution had their beginnings as a result of rapid technological advances at the end of the World War II (1945). These advances in transportation, industry, and agriculture made life easier for us, but at the same time the environment began to be damaged to a greater extent. Some of these products that caused pollution are auto-

Figure 7.2: OLD CAR WITH INEFFICIENT ENGINE. Before catalytic converters, car engines produced damaging exhaust to a greater degree than today's engines.

mobile engines, sewage treatment plants, and the "wonder material," plastic.

Automobiles are very useful; we'd hardly know how to live without them; however, over the years we've had to deal with automobiles and their effects on our Earth. As time passed, engines became more powerful and more polluting. Newer engines put larger amounts of nitrogen oxide gases into the engines' exhausts. **Catalytic converters**, an anti-pollution device in today's cars, removes most pollutants that automobile engines produce. As far as automobile engines are concerned, we have a lot to do before they are pollutant-free.

Another technological development that was meant to protect the environment but added to the pollution problem was the invention of **sewage treatment plants**. Plant and animal matter is removed by the use of bacteria and oxygen in breaking down organic matter and turning it into inorganic nutrients. Next, the nutrients are added to the water and, unfortunately, algae growth is increased. The increased growth of algae adds more organic matter to water, depleting the water cycle of

Figure 7.3: TRAFFIC. An everyday event, millions of cars travel our highways, polluting our environment. (Warren Gretz—NREL)

oxygen. Today sewage research is working on methods to remove the inorganic nutrients as well.

Plastic, the "wonder material", is used across the world. Plastics are used in schools, churches and industries. Some plastic products are found in cars, appliances, shoes, kitchens, workshops, toys, industries and hospitals. They are very useful and durable. They are so long-lasting that they don't break down and cannot be reabsorbed into the soil.

Through the production of plastics, pollutants are formed as well, as large amounts of energy are needed for their production. This demand for more electrical power causes more fuel to be burned which in turn causes more pollution. An ongoing chain of events helps reduce pollution in one area and helps increase pollution in another.

In 1899, the U.S. government passed its first pollution control law. It prevented liquid waste, with the exception of sewer water, from being dumped into navigable water. This law was not enforced to any degree, and therefore was ineffective. In the 1960's, pollution laws were passed, and by 1970, fifteen federal pro-

Figure 7.4: PLASTICS CAN BE RECYCLED. Durable plastics, called the "wonder material" of the 20th century, actually caused a major pollution problem because of their nearly non-biodegradable characteristic.

grams were united by the Environment Protection Agency (EPA). The EPA reports to the President of the United States. It has the power to enforce laws passed, set and enforce standards, conduct research, and assist government at all levels.

We have caused several kinds of environmental pollution: air, water, and soil pollution. Our surrounding environment is involved with so many relationships that one form of pollution exerts its effects on many organisms. Water pollution harms or kills the organisms that are living in that environment and reduces the food available for other species. As food availability is reduced, some species are wiped out, while others barely survive. Wind blows land pollutants into nearby waterways, into the atmosphere, and to other parts of the land. All parts of our environment eventually become contaminated.

Our government has passed laws to reduce pollution by forcing businesses and individuals to stop or curtail polluting activities. As citizens of this country and inhabitants of Earth, each of us must stay aware of polluting situations.

Look in your local newspaper for a pollution article. Cut out, read, and give a summary of the article. What is your reaction, your opinion, and your solution to this problem? Staple to this page.

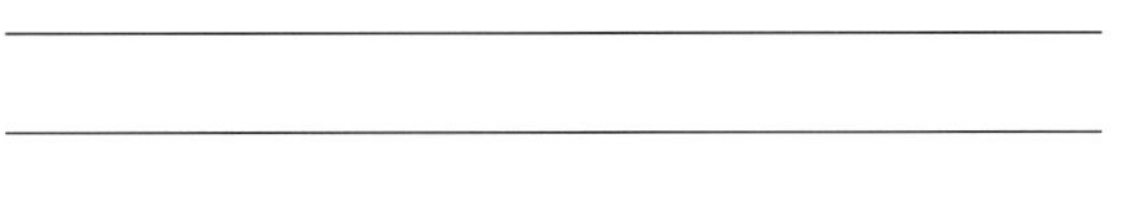

Air Pollution

The first form of pollution we will discuss is **air pollution**. Pollutants that enter the air change the colorless, odorless, combination of gases we call air into a grayish, foul-smelling, smog (in heavily polluted areas). This polluted air is dangerous to breathe and may cause many health problems. It also has the potential to kill plants and even destroy property.

By releasing many millions of tons of dangerous gases and other **pollutants** into the atmosphere, we change clean air into polluted air. Pollutants such as carbon monoxide, smoke, nitrogen oxides, ozone, and sulfur di-

Figure 7.5: CURRENT EVENTS. Many newspapers print local and national articles on pollution for consumer awareness. Most newsprint is partially composed of recycled paper.

oxide pour into the air from an array of sources. Such sources are airplanes, cigarette smoke, car, truck, and bus exhaust, and the burning of coal as an energy source in factories and homes.

The process of burning a substance for power and energy combustion results in the release of pollutants. These pollutants are poisonous and may range in color from clear white to a thick, heavy, black smoke. Combustion of oil, gasoline, and coal makes up about 85% of these pollutants. Fossil-fueled electric power plants are responsible for much of the nitrogen oxides and pollutants emitted into our atmosphere. About 95% of carbon monoxide pollution comes from the combustion of gasoline and diesel fuel in motor vehicles. Catalytic converters are now used on automobiles to reduce pollution produced from the combustion of gasoline. Other major producers are pictured in Figure 7.6.

When these waste gases mix with smoke and fog, smog results. **Smog**, a thick layer of polluted air, can be seen lying near highly industrialized cities. When sunlight acts on waste material in the air, photochemical smog is formed. **Ozone (O_3)** is one of the products of this action and irritates the lining of the nose and throat and causes eyes to sting and water. A lot of photochemical smog is found in Southwestern United States due to the lack of wind. Smog can build up to dangerous levels which can cause sickness and death.

Both smoke and airborne **particulates** (tiny particles) may have harmful effects on human health. Eyes and lungs may become irritated if the particulated matter settles into the lungs, and serious respiratory diseases may occur. Bronchitis, asthma, allergies, pneumonia, emphysema, and now, even cancer, are thought to be caused by airborne pollutants.

Black lung is a disease caused by inhaling coal dust from coal mines. Miners spending long hours, year after year in coal mines may develop black lung due to the coal dust irritating the lung tissue. **White lung**, a disease caused by inhaling dust from glass manufacturing, is another illness created by our industrial nation. Chest x-rays can detect these lung diseases.

Air pollutants may be reduced or built up by weather conditions. Wind may blow the pollutants around while rain and snow wash them into the soil, reducing air pollution levels. Pollutants are usually put into the air at a greater rate than are removed by weather. In addition, there are times when **thermal inversion** occurs and air pollution levels are built up. In thermal inversion a layer of warm air settles above a layer of colder air and holds this cold air mass close to the ground. The air pollutants are prevented from scattering and remain concentrated closer to the Earth. Seriously high levels of pollutants occur in densely populated cities or highly industrialized areas.

It is estimated that only 3 days of temperature inversion or poor mixing of our air leads to hazardous conditions in highly populated areas. History shows death may occur during times of inversions. In London thermal inversions in 1552, and again in 1962, killed 4,000 and 700 people respectively. A well-publicized inversion occurred in 1984, in Bhopal, India, killing 3,300 people and causing 20,000 people to become ill. Air pollutants released into the air and kept close to the Earth's surface by the weather were the cause of these deaths.

Figure 7.6: AIR POLLUTION. Air pollution sources vary. The top photo shows a sugar cane refinery pouring pollutants into the air while, in the bottom photo, an electricity generating power plant produces both air and water pollution.

Increased air pollution levels caused by many factors exhibit their effects on plants as well as nonliving materials. Poisonous gases in the air may restrict plant growth and eventually prove fatal. Vegetable gardens in New Jersey, forests in Tennessee, and citrus groves near Los Angeles have all been damaged by air pollution.

The reduction of or total destruction of crops in populated areas due to air pollution may exert its effects nationwide or even worldwide. Citrus crops that were damaged in California caused shortages and high prices on the East Coast of the United States as well as in nearby towns.

Our homes, cars, and public places become dirty and worn due to the negative effects of air pollutants. Even steel and concrete are affected by pollutants. Many of our treasured statues worldwide have been removed and placed inside buildings to protect them from further destruction.

Our climate is affected by air pollutants as well. Gases and other pollutants can change the temperature of an area by reducing the amount of sunlight that actually reaches the ground. Carbon dioxide found in the air causes sunlight that has reached the land to remain in the atmosphere longer than usual. The greenhouse effect is the warming resulting from the heat staying in our atmosphere for a longer period of time instead of being reflected back into space. As the heat-trapping gases become more concentrated from air pollution, the average temperature continues to rise.

UV INDEX

The UV index number registers the amount of **u**ltraviolet radiation reaching the Earth for one hour, around noon. The higher the UV number, the higher the amount of radiation, and the greater risk of sunburn and eye damage. UV number is **not** linked to the SPF factor on sunscreen products.

Higher UV numbers are found:

- at higher elevations;
- near the tropics;
- from water reflection, as well as snow and ice;
- near holes in the ozone layer, or thinning layers;
- on sunny days.

Figure 7.7: CROP DAMAGE. Pollution has caused damage to citrus crops nationwide resulting in shortages and higher priced fruit.

Figure 7.8: JEFFERSON MEMORIAL, WASHINGTON, D.C. Most of our buildings show the effects of air pollution and need to be refurbished periodically, a costly maintenance bill to the taxpayers.

Destruction of the **ozone layer** is another major problem caused by pollution. Our ozone layer is a layer of gas in our atmosphere that protects plants and animals from harmful ultraviolet light. The ozone layer ranges from 18 to 19 km (12-30 miles) above the surface of the Earth. It forms naturally by the action of the sun's energy on oxygen.

Scientists in the 1970's identified substances called chlorofluorocarbons (CFC's) in aerosol sprays and refrigerants. The use of CFC's in aerosol cans has been banned in the United States. CFC's have the ability to thin the ozone layer. This thinning of the ozone allows more of the ultraviolet radiation to reach the Earth endangering mankind's health. In some areas, especially in Antarctica, a loss of ozone was detected. Recent evidence shows the "hole" fluctuates seasonally. It is larger in the spring and stays that way for several months. After this period of time, the "hole" shrinks, but the actual level of ozone is decreasing. A similar situation has been detected over the Arctic region.

The United States and 36 other nations started protecting the ozone layer in 1987. A few years later in 1991, an upper atmosphere research satellite was launched by NASA to monitor ozone depletion on a worldwide level.

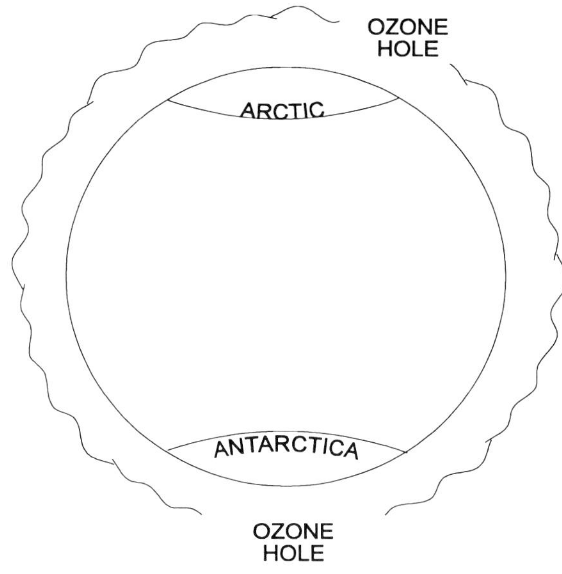

Figure 7.9: OZONE LAYER. Scientists have observed openings in the ozone layer above both polar regions. The openings fluctuate seasonally.

As the countries around the globe work towards minimizing the depletion of the ozone layer, we must protect ourselves now. One protection method is simply to use sunscreen as a skin protective device. A good portion of the UV rays are blocked out by sunscreen which has an SP factor of 15 or more. Another protection method is to cover up while in direct sunlight during the hottest part of the day, 10 AM to 2 PM. We may all desire a golden tan, but covering up and using sunscreen may prevent the suffering associated with skin cancer.

Our government passed the **Clean Air Act in 1967** that established the standard set for air pollution control. The **Environmental Protection Agency (EPA)** carries out the requirements of this act. Standards in concentrations of hazardous substances are specified as well as discharge of pollutants into the air. An amendment in 1990 listed carbon monoxide, ozone, particulate matter, acid rain and air toxins as major air pollutants.

As an air pollutant, **acid rain** is a current topic of discussion. Acid rain is any form of precipitation that has been polluted by acids. Nitric and sulfuric acids are commonly found in acid rain. Compounds from the combustion of gas, coal and oil form sulfur dioxide and

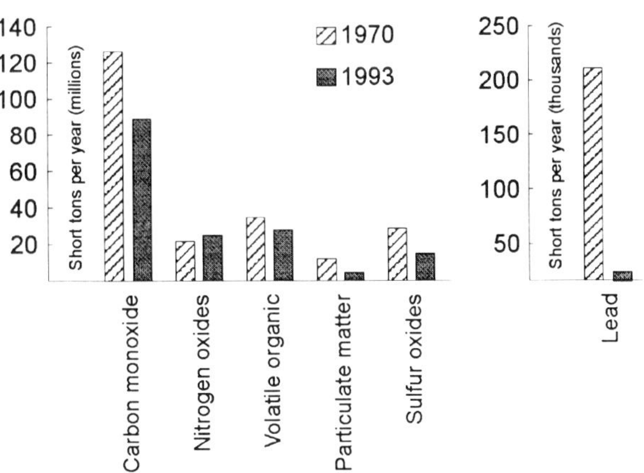

Figure 7.10: AIR QUALITY IMPROVEMENT. The Clean Air Act of 1970 has helped reduce emissions of several primary pollutants.

nitrogen oxides. These compounds react with water vapor in the air to form acid rain. Thousands of waterways, including lakes, rivers, and streams have been harmed. There the fish and other aquatic life die due to the change of the level of acid in the water. In some cases the pH level (see Figure 7.11A), relative acidity, has changed so drastically that entire populations of fish have been destroyed. Scientists believe acid rain also destroys buildings, statues, forests, bridges, and soil. Many parts of the world are affected by acid rain including North America, parts of Asia and Europe. Rural areas are affected by acid rain as well, due to taller smoke stacks in urban areas sending smoke and pollutants higher into the air and winds carrying the pollutants farther away from the source (industrial complexes or cities).

Several ways of reducing acid rain include devices which reduce nitrogen and sulfur compounds from fuel or emissions before they are released into the atmosphere. We now know that adding lime (base) to waterways and drainage areas temporarily neutralizes their acidity, but neutralization may have harmful effects on other organisms and matter.

There have been many ways we have reduced air pollution:

- We have placed filtering devices within smokestacks to filter out solid particles formed when fuels are burned.

- We have reduced waste gases by more complete burning of fuels.

- We have replaced coal and oil with natural gas (CH_4) since it burns more completely than oil and coal.

- We have used oil and coal that is low in sulfur or have had sulfur removed before it's burned.

- We have pursued alternatives to gas engines in our vehicles:
  - natural gas or hydrogen engines
  - turbine engines
  - electric cars using rechargeable batteries.

In every way possible we must reduce what goes into our air, our most necessary resource; our lives depend upon it.

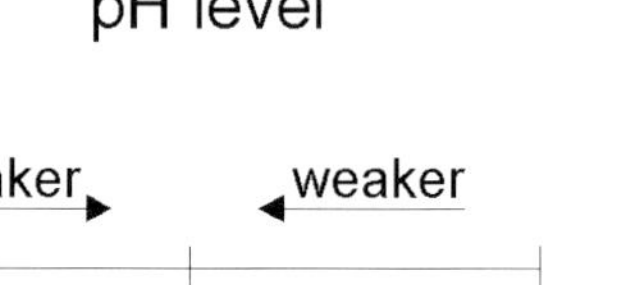
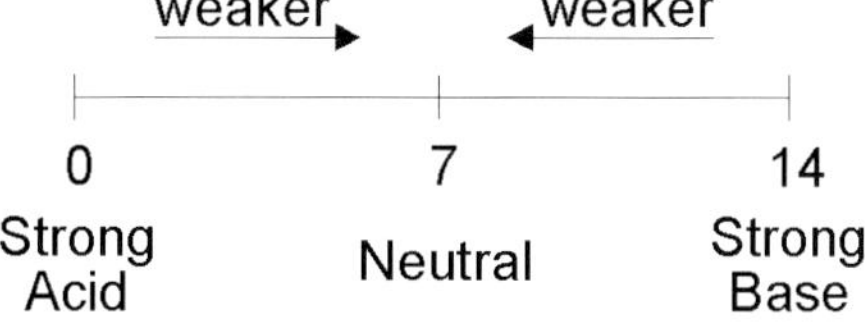

Figure 7.11A: pH SCALE. Strong acids are close to the zero point. Acid rain lowers the pH of water in many ponds, rivers, and lakes which kills many species of aquatic life.

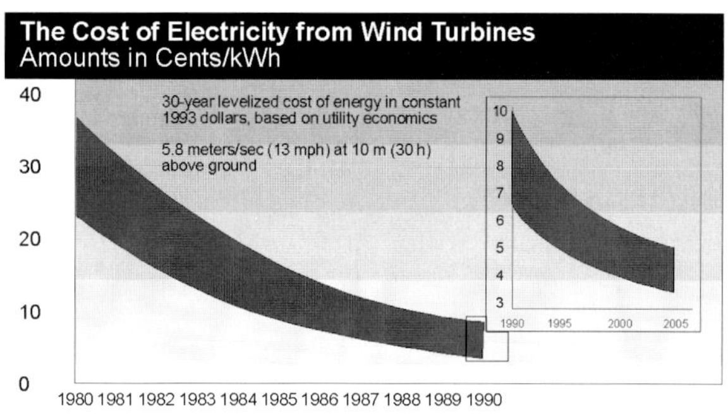

Figure 7.11B: WIND POWER. We now have the ability to generate electricity by wind power which is clean, efficient, and less costly. (U.S. Department of Energy)

↘ DID YOU KNOW?

Indoor air pollution has been studied since the early 1980's. The main indoor pollutants are:

1. Tobacco smoke

2. Household chemicals (cleaning supplies)

3. Carbon monoxide from furnaces, gas stoves, garages, etc.

4. Formaldehyde (in furniture, paneling, fabrics, permanent-pressed clothes, carpet, tooth-paste, paper products, medicines, etc.)

5. Asbestos (still found in insulation in some homes and offices.)

6. Radon, a naturally occurring gas in the Earth's crust that may leak through basements and pipes.

✔ THINGS TO KNOW AND DO:

Define:

* air pollution ____

____

____

*indoor air pollution ____

____

____

*pollutants ____

____

____

*catalytic converter ____

____

____

*smog ____

____

____

*thermal inversion ____

____

____

*CFC's ____

____

____

*EPA ____

____

____

*acid rain ____

____

____

*radon ____

____

____

✍ FILL IN THE BLANKS:

1. Air pollutants include ________ , ____________ ,
   ____________ , ____________ , and ____________
   ____________ .

2. ____________ , ____________ , and ____________ are
   three sources of air pollution.

3. ____________ ____________ reduces pollution from automobile exhaust.

4. Smoke + fog = ____________ .

5. ____________ ____________ and ____________
   ____________ are two diseases caused by inhaling job-related pollutants.

➔ MODIFIED TRUE-FALSE: correct underlined word if necessary.

________ 1. Chest x-rays <u>cannot</u> detect lung diseases. ____________

________ 2. Wind may <u>reduce</u> air pollution levels. ____________

________ 3. Thermal inversion occurs when a layer of warm air settles <u>below</u> a layer of cold
   air. ____________

________ 4. Climates <u>are</u> affected by air pollutants. ____________

________ 5. Ozone holes are located above the <u>equator</u>. ____________

MATCHING:

________ 1.   The Clean Air Act of 1967.   a. killing 3,300

________ 2.   Amendment, 1990   b. standard set for air pollution control

________ 3.   Thermal inversion   c. list of major air pollutants
              (London, England)

________ 4.   Thermal inversion   d. killing 4,000
              (Bhopal, India)

✎ IN SEARCH OF...

In order to learn more about local pollution, call a manufacturing company nearby and document the following:

1.   Company ______________________

2.   Name of person you spoke with ______________________

3.   Date ____________________

4.   Phone number ____________________

5.   How do they **control** air pollutants? (what methods, what pollutants, etc.)

____

____

____

or

Make a comprehensive list of indoor pollutants. (at least 50 **numbered** items).

Water Pollution

Pollutants in water cause our water supply to be unfit for human consumption. Drinking, cooking, bathing, or watering our plants and livestock all require good quality water. **Potable** (drinkable) water in some parts of the world is a rare resource. There are places in our world that do not have proper sewage, proper irrigation, or a clean water supply. Sickness and diseases arise from these areas of the world because of the lack of potable water.

The main resources of **water pollution** come from industries, farms, and sewage treatment systems. Oils, gases, untreated sewage and various chemicals and minerals, pesti-

According to the EPA . . .

- 30% of U.S. rivers, 42% of lakes, and 32% of estuaries are still being polluted by silt and nutrients from farm and urban runoff.

- sewer overflows and municipal sewage contribute to the pollution of our waterways.

- 12% increase of toxic chemical discharged into surface water

- farm run-off and storm water account for 1/2 of all pollutants in U.S. surface waters.

- only 200 of 1200 priority sites have been cleaned up since 1980.

cides, fertilizers, as well as wastes from plant and animal matter are some of the pollutants that are dumped in huge quantities into our water supply daily. Even when wastes are dumped onto the land, especially hazardous waste from nuclear power plants, some of it may find its way into the water supply, increasing the water pollution level. In addition, pollution reduces the oxygen content of water. The lower level of oxygen causes more pollution by killing fish, plants, and other aquatic organisms that need oxygen to live.

Sewage dumped into the water causes many health problems. Sewage treatment systems help reduce the harmful material in **effluent**, (the waste flowing from sewage treatment plants) Even when sewage is treated, some of the materials left still harm our water supply. In our modern treatment plants, sewage is stored in tanks where it is constantly mixed with air. Oxygen in the air combines with the harmful substance found in sewage, making it less harmful. Now, this treated sewage does not remove the needed oxygen required by fish and other organisms.

There are natural **cycles** existing that are able to absorb some waste. **Aerobic bacteria**, agents of decay, work on wastes such as dead fish and break it down into nutrients, phosphates, and CO_2. The nutrients are used by algae, which serve as food for the **zooplankton**, microscopic

Figure 7.11C: TO SWIM OR NOT TO SWIM? Sewage dumped into our water causes many health problems. Sewage treatment plants take care of most of this, but we need to do our part in cleaning up our waterways.

Figure 7.12: GROUND-WATER CONTAMINATION. A drilling rig installs a monitoring well to evaluate ground-water contamination at a chemical facility. Consultant on the left is monitoring drill cuttings from the bore hole for volatile organics such as gasoline.

water organisms. The cycle continues as small fish feed on the zooplankton and so on. Now, if too much waste is put into the waterways, the aerobic bacteria uses too much oxygen which, in turn, reduces the fish population, and the water becomes dirtier and dirtier.

Other situations exist in which fish and plants are killed; one of these is **thermal pollution** caused when heated water is dumped into ponds, lakes, and rivers. The heated water raises the temperature of the body of water to a temperature that kills fish. The water may come from electric power plants and industrial plants that use water to cool their machinery. The dead fish cause more oxygen to be removed by the aerobic bacteria and once again the cycle is interrupted and polluted water results.

Polluted water may also result from oil spills. Oil drilling platforms and oil tankers occasionally have accidents, spilling millions of gallons of oil into our oceans. Besides ruining our beaches and killing marine life, birds and other mammals are hurt or killed, as well.

The birds' feathers become soaked with thick, black oil making it impossible to fly. Oil getting into the lungs of mammals gives little chance of survival for the animal. One such oil spill occurred in North American waters in 1989. A tanker hit a reef near the port of Valdez, Alaska, spilling about 11 million gallons of crude oil into Prince William Sound. During the Persian Gulf War in 1991, Iraq released an estimated 465 million gallons of oil into the Persian Gulf, polluting the waters and killing wildlife. During that same war, Iraq set about 650 oil wells on fire, polluting the air over Kuwait. The oil wells burned for a long time and even caused total darkness during daylight hours. The last of the oil well fires was extinguished in November, 1991. Other oil spills have occurred worldwide and have been disastrous, to both our water supply and marine life.

Water may also become overly enriched by nutrients added to water through **eutrophication**, nutrient enrichment. Nitrates and phosphates from fertilizers and detergents respec-

Figure 7.13: DRILLING WITH MUD. These sophisticated drills recirculate mud to cool down the drill bit and to pack the bore hole to keep it from collapsing. Drills like this one install monitory wells which allow scientists to test groundwater (even through bare rock). Underground storage tanks at gas stations are a major source of contamination nation-wide. Double-walled tanks and piping with leak detection are now replacing older, single-walled steel storage tanks and pipes.

News Flash!

EPA studies show that 116 million Americans drank contaminated water in 1992-93.

Another News Flash!

According to a more recent study, 1994, 14.1 million Americans regularly drink water contaminated with five major agricultural herbicides. These contaminates are now linked to birth defects, bladder cancer, and endocrine system disrupters.

News Flash #3

48 toxic chemicals were listed as priority phaseouts because of their perilous threat to people and wildlife. (National Wildlife Federation, 1994)

---

tively cause rapid algae growth in water. With this increased algae growth, comes increased algae decay, using more and more oxygen, thereby depleting the water's oxygen supply. As the oxygen level is lowered, other aquatic organisms perish.

Water pollution from various sources may cause a number of diseases to spread. Dysentery, cholera, typhoid fever and other similar diseases are easily spread by polluted waterways. It has been estimated that about 75% of our nation's community water supplies are disinfected with chlorine to kill many disease-producing organisms. Heavy metals such as mercury, lead, polychlorinated biphenyls (PCB's), chloroform, and arsenic, are not removed by the disinfection process **chlorination**. Our ground water supplies are in danger of being contaminated with such chemicals and metals resulting from the careless placing of toxic waste in dumps. Scientists suggest that even tiny amounts of these waste products consumed by humans over long periods of time may lead to serious health problems.

The U.S. Congress in 1974 passed the **Safe Drinking Water Act** in order to keep the nation's water supply safe. The EPA set the standards for water quality and in 1977, along with the state governments, began enforcing its standards. The standards set would keep the amount of harmful bacteria, chemicals and metals in drinking water at a reduced "safe" level.

Guidelines to limit chloroform and other trihalomethanes (THM) levels were issued by the EPA in 1979. THM's are found in drinking water in large cities due to the addition of chlorine at water treatment plants. It has been stated that exposure to THM's is thought to increase cancer risk.

Nations have begun to work with one another in trying to solve water pollution. More than 70 countries have gathered to put their heads together to improve water treatment methods, share knowledge on desalination and to come up with more efficient ways of using and conserving the world's water supply.

Some methods of water treatment are as follows:

1. coagulation and settling
2. filtration/chlorination
3. aeration
4. water softening
5. desalination

Coagulation and Settling

Water to be treated is mixed with **coagulants**, chemicals that form sticky globs called **flocs**. The coagulants commonly used are alum or aluminum sulfate. The impurities in the water such as bacteria and mud stick to the flocs. The water is then passed into a settling tank, where the flocs and impurities are later removed from the bottom of the tank.

Filtration/Chlorination

Water passes through a filter consisting of sand, coal and gravel. As the water trickles through the filter, impurities are removed and are now ready to flow to reservoirs for **chlorination** that kills the bacteria. Chlorine may be added to the water either before or after filtration, or, in some cases, chlorination is the only water treatment some cities use.

Figure 7.14: AERATION. Fountains such as this one spraying water into the air remove odors and put oxygen into the water.

Aeration

Some processes of water treatment remove unpleasant odors and tastes from the water. The **aeration** process sprays water into the air. The oxygen in the air removes the bad taste and odor.

In some areas water softening contains minerals making **"hard" water**. Hard water does not lather well and forms deposits on pipes, equipment, and bathroom tile. To reduce the mineral content (changing the hard water to soft water), lime or soda ash is added to the water. Activated carbon is used also to improve taste and odor and to remove chemicals from the water.

"Look, Ma—no cavities"

Fluoride is added to water in most states and has been effective in reducing tooth decay.

Desalination

Desalination (taking salt from saltwater) includes distillation, reverse osmosis, electrolysis, and freezing. **Distillation** involves heating and condensing salt water vapor to produce pure water. **Reverse osmosis** uses high pressure causing fresh water to be squeezed through a membrane, leaving the salt behind. **Electrodialysis** is based on the fact that when salt is dissolved in water, salt breaks into ions, positive and negative. Two types of membranes are used in this process. An electric current is sent through the water; positive ions pass through one membrane and the negative through the other; the salt is then drawn off, leaving fresh water. The **freezing process** produces pure water ice crystals from salt water. The salt is separated and trapped from the fresh water ice crystals. This method is very expensive and is not used very often.

Much money has been spent to preserve and treat our water supply. As good stewards we must learn to guard this valuable resource carefully, use it wisely, and prevent its deterioration. Remember, "man does not live by bread alone...."; he needs a little water to wash it down.

Figure 7.15: ELECTRODIALYSIS. This water plant on Sanibel Island, Florida uses this process to produce 4 million gallons of fresh water a day.

From Salt Water...

In Cape Coral, Fla., a reverse-osmosis desalination plant produces about 14 million gallons of water a day.

↘ DID YOU KNOW?

- 37 billion gallons of H_2O a day is supplied to the US population by municipal water departments.
- We use an estimated 100 gallons of H_2O daily per person at home.
- It takes 7 and 1/2 gallons of water each time you "flush."
- 70% of the world's fresh water is found in Antarctica.

✔ THINGS TO KNOW AND DO:

Define:

*potable ____

____

____

*water pollution ____

____

____

*aerobic bacteria ____

____

____

*zooplankton ____

____

____

*thermal pollution ____

____

____

*eutrophication ____

____

____

*coagulation and settling __________________________________

____

____

*aeration ____

____

____

*desalination ____

____

____

*chlorination ____

____

____

List at least 7 pollutants that are dumped into our water supply.

1. _________________________   4. _________________________

2. _________________________   5. _________________________

3. _________________________   6. _________________________

7. _________________________

List and explain two tragic oil spills that have occurred since 1989.

1. Place(s) _________________________

____

2. What happened: _________________________

____

____

____

____

3. What did this do to birds? _________________________

4. Mammals? _________________________

5. Beaches? _________________________

➡ MODIFIED TRUE-FALSE: correct underlined word if necessary.

________ 1. U.S. Congress passed the Safe Drinking Water Act in <u>1970</u>.

____

________ 2. Limiting chloroform and THM levels was issued by the EPA in <u>1974</u>.

____

________ 3. Fluoride <u>promotes</u> tooth decay. _________________

________ 4. Alum and aluminum sulfate is used in the <u>filtration</u> process.

____

________ 5. Flocs <u>attract</u> impurities in the water. _________________

________ 6. Outdoor lake fountains are examples of aeration. _________________

________ 7. Desalination includes <u>forward</u> osmosis. _________________

________ 8. Distillation <u>requires</u> condensation. _________________

________ 9. Eutrophication <u>improves</u> water quality. _________________

________ 10. <u>Chlorination</u> may be the only water treatment some cities use.

____

Soil Pollution

The outer layer of the Earth's crust has a thin covering of fertile soil needed by living organisms. The world's food is grown in this fertile layer, and it also provides food (fields) for livestock and wildlife. **Soil pollution** occurs when soil is contaminated by the addition of pesticides and by fertilizers or other foreign materials.

Soil takes thousands of years to form and just a few years to be destroyed. Fertilizers added to the soil and pesticides sprayed on crops, both used in hopes of better crop yield, harm nature's soil cycle.

Nature's cycle helps keep the soil fertile. The waste of flora and fauna and dead organic matter find its way into the soil. There, in the soil, bacteria and fungi (agents of decay) break down this matter into nitrates, phosphates, and other needed nutrients. The producers (green plants) use these nutrients in their life cycle and, in turn, make food for consumers. Fertilizers and pesticides added to the soil may decrease the ability of bacteria to decay waste and, hence, to produce nutrients. We must be careful of what we do to the soil because what

---

Organic Gardening

- Composts and manures enrich the soil without the use of inorganic chemicals, pesticides or fertilizers.

- Natural insecticides made from garlic juice are used to repel insects.

- Beneficial insects like ladybugs eat plant aphids, food insects that harm many crops.

- Planting marigolds between rows of vegetables keeps infestation of harmful insects down and may keep rabbits from "dining."

---

we do may have negative effects that harm bacteria and other useful soil organisms. What could a farmer do to harvest good crops without the use of pesticides and fertilizers?

____

____

____

Figure 7.16: ABUNDANT FLOWERS. Producers such as these require nutrients in their life cycle that they obtain from the soil. We must maintain good soil quality so our producers can continue to produce in abundance.

Solid Waste

Probably the most visible form of pollution is **solid waste**. This form of pollution affects our air, water and soil. Litter along roadsides, on beaches, parks, floating in oceans, ponds and lakes or heaped up in huge mountains in dumps or in landfills are examples of solid waste. Solid waste comes in all sizes, shapes and colors. Paper, scrap metal, plastic, cars, appliances, slag from mining and tires are some examples of solid waste. Can you think of more? What kinds of solid waste do you see in your city or town?

____

____

____

Where are these wastes located?

____

____

____

Figure 7.17: SOLID WASTES. Junk yards contain many forms of solid waste like these cars while beaches and waterways collect their own form of solid waste. Volunteers help with the clean-up.

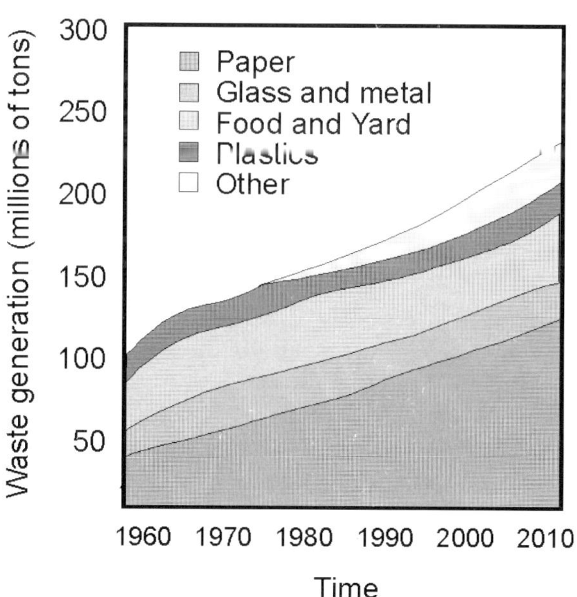

Figure 7.18: SOLID WASTE GENERATION, 1960-2010. As you can see, solid waste keeps increasing yearly and is projected to continue to rise. (Biosphere 2000, p. 443)

Where to put solid waste is a big issue with which our world has to deal. In the past, oceans and unpopulated land areas, for example, have been our trash bins. This disposing of solid waste in such a careless manner has led to the serious state of pollution we are facing today. There may be some damage to our environment wherever we choose to dispose our waste.

> Hazardous waste sites 1,235 in U.S.A. (EPA, 1992)

Dumps provide places for disease carrying animals such as rats to live. Dumps are also unattractive. **Landfills**, large open areas, contain most of our solid wastes; however, space for landfills and dumps is decreasing as solid waste is increasing.

By the mid 1980's , more than 2 billion short tons* of solid waste was being produced yearly in the United States. This waste includes agricultural, industrial, household, and mining waste. Landfills are the #1 spot for getting rid of waste. Now, we are finding more and more waste that is difficult to dispose of completely. Aluminum is replacing tin and steel cans; however, aluminum stays in its original form for many years. In contrast, steel and tin cans break down through oxidation (rusting) and are absorbed by the soil. Paper and cardboard have been replaced by plastic and the same situation has happened; plastic does not decay, and it gives off harmful gases when burned.

Today solid waste is an environmental problem that affects us all. In one way or another, all creatures, not only man, contribute to the increase of solid waste. Man has definitely upset nature's balance by his development of "new and better products" without considering the impact these may have on polluting the Earth. Let's join together to reduce solid waste by: using products that decay in a timely manner, reducing the amounts of materials we use, and reusing items for other purposes.

Figure 7.19: NUCLEAR ENERGY. This is a cooling tower at Davis-Besse Nuclear Power Station near Port Clinton, Ohio. (Biosphere 2000—p. 237)

> The amount of waste generated per person, per day in the United States is approximately 4.4 pounds. (EPA, 1995)

Other Kinds of Pollutants

Other pollutants such as mercury, lead, radiation, and noise exert their effects on our world.

Noise pollution from airplanes, music, construction and industry, for example, cause discomfort to many people and animals. In cases where the noise level is extreme, hearing damage and even deafness may occur. Other diseases such as nervous conditions, high blood pressure and ulcers have been linked to excessive noise.

> One decibel is the smallest difference between sounds detectable by the human ear; 120 decibel sound is painful.
>
> **Decibels**
> 30—normal conversation
> 70—normal traffic
> 90—heavy traffic
> 100—jet plane at takeoff
> Rock music is 80 decibels.

Electromagnetic radiation consists of electromagnetic rays produced from x-ray machines, TV sets, microwave ovens, lasers and computers. Large amounts of exposure can cause cancer, but scientists are not sure what small amounts of exposure to humans may cause. Radiation from nuclear weapon testing and nuclear reactors contributes to our existing pollution problem. Today most nuclear weapon testing has been banned internationally. Reducing and storing of the radioactive waste is currently being studied by scientists.

We have mentioned acid rain and pesticides as pollutants earlier in this chapter and now we must consider **mercury and lead** as pollutants to our environment and threats to our lives. In

---

*   2,000 pounds (short ton) 2,240 pounds in British weight (long ton)

the form of liquid and gas, mercury and lead (heavy metals) are being dumped into our air and water. Most of the pollutants come from motor vehicle engines and the combustion process of industries; these pollutants are long-lasting and may contaminate large areas. They are able to collect in tissues and organs of organisms and are passed through the food chain. They are highly poisonous and can damage the nervous system. Dangerous levels of mercury and lead have in the past been noted. We were warned at one time, not to consume tuna fish due to elevated mercury levels. Lead in paint had caused brain damage to children who were eating wall paint that had chipped off. Today, most house paint is lead-free.

We have put a lot of problem areas on hold by introducing new products promised to be less toxic, non-toxic, or less damaging to the environment. Still we are left with the problems of the past haunting our future. All of us, including you, must commit some time and energy to slow down pollution in any form. Don't wait for the other guy to help because he's probably waiting for your lead. As good stewards, let's begin the task at hand.

Figure 7.20: UNLEADED GAS. In an attempt to decrease the amount of lead being dumped into our air, today's cars now require only unleaded fuel.

> Today's blood lead levels of children have been reduced by 77% since 1981

↘ DID YOU KNOW?

- In 1993, millions of tons of sand and other debris were left on crop land by floodwaters, especially around the Missouri and Mississippi Rivers. This pollution severely damaged farms and wetlands.

✔ THINGS TO KNOW AND DO:

Define:

* soil pollution ____

____

____

*organic gardening ____

____

____

*landfills ____

____

____

*noise pollution ____

____

____

*electromagnetic radiation ____

____

____

*mercury and lead pollutants ____

____

____

*sewage treatment plants ____

____

____

*plastics ____

____

____

Name at least 5 places where solid waste is found.

1. _______________________   4. _______________________

2. _______________________   5. _______________________

3. _______________________

Explain in at least 3 sentences, how organic gardening reduces soil pollution.

____

____

____

List 5 examples of solid waste.

1. _________________________________ 4. _________________________________

2. _________________________________ 5. _________________________________

3. _________________________________

What problems today do we have with landfills?

____

____

____

How are newer and better products linked to solid-waste pollution?

____

____

What 5 items produce electromagnetic radiation?

1. _________________________________ 4. _________________________________

2. _________________________________ 5. _________________________________

3. _________________________________

What is one harmful effect of this form of pollution?

____

____

RECYCLING

As pollution increases, anti-pollution research increases, but only small reductions in pollution levels result. Our increased desire for convenience has caused much of our pollution problem. We are the "throw away and over-packaging" generation. Much of the packaging of soda, canned goods, cereals, appliances, etc. is unnecessary and is thrown away.

Currently, recycling programs and education for citizens have been set into motion. We are now collecting, sorting, saving, reusing, reducing, and recycling many waste products and waste containers.

Recycling is the collecting and processing of waste material for new products. Our waste in the past has been put together and dumped into trash bins and collected by our town or

Figure 7.21: IGGIES ON POLLUTION PATROL. A troupe of "Iggies" are waiting to be fed with plastics, aluminum, oil, newspapers, white paper, glass, etc. Will you donate something?

Figure 7.22: RECYCLING. Curbside recycling today is not always able to pick up all recyclable materials. "Iggies" such as this one will accept some of these uncollectibles.

city. This mixing of all types of wastes, as well as recyclables, reduces the chance and value of recycling these materials. Today we sort out recyclable wastes and place each type into different collection containers, keeping the items clean and easy to sort and therefore more valuable to a recycling program.

Steel and aluminum cans are collected at many residences but can also be bagged and dropped off at collection bins. Recycling plants get the materials next where recycled steel and aluminum can be used to make new cans.

Glass and plastic containers are marked with recycling stamps. Glass containers can be refilled or melted down and reused. As we become more and more accustomed to recycling, the need for solid waste disposal and landfills is reduced. The litter along roads and waterways disappear as people become more conscientious about pollution and its consequences.

Many types of waste can be recycled. We mentioned cans, glass, and paper, but other products are recycled as well. Old tires are melted down, chopped up, and cut into strips

for many uses. Gases, chemicals, and oil are also collected through recycling. Old newspapers are turned into pulp, then, into clean newsprint. Both glass and tires can be recycled into road building materials.

Many materials are being recycled through government actions as well as by private organizations. Our government, federal, state, and local, is trying to control pollution by passing laws and by adding recycling programs. State and local agencies have developed their own recycling programs. These are very different and are custom-tailored to fit particular areas. The most common program is the curbside pick-up of recycled materials such as aluminum, newspaper, plastic, and glass from our homes.

In 1992, over 5400 U.S. communities provided curbside recycling programs. This type of program makes it convenient for the individual, but sometimes it can be expensive. Curbside collection requires paid labor, vehicles, insurance, and maintenance for the vehicles, etc. The expense is offset by a participation rate that citizens pay and by revenue generated from the sale of materials collected. Curbside recyclable goods are easy pickings for individuals who steal for their own profit, thus reducing revenues for our cities.

Economics encourages recycling programs. Recycling prolongs the life of existing landfills, which delays the need of new landfill sites. This comes down to money saved. If you or I were charged the "real" cost of waste disposal, we would be more careful in our purchases and in our throwouts.

Several states have mandatory recycling laws. Pennsylvania and New Jersey are two states that require consumers to recycle goods and materials. Recyclable materials are collected and stored in some cases for future use.

The need for recycling laws has several reasons. Manufacturers do not want to invest time and money to redo their plant or have an increase in their cost of production. Mining and logging companies are given tax deductions and subsidies for complying with recycling laws; this helps keep their cost of goods low.

Another law creates markets for recyclable materials. Thirteen states require newspaper

Figure 7.23: SALVAGE. Many businesses generate lots of solid waste. Collection and recycling of these materials reduces our need for additional land fills.

companies to use a certain amount of recycled newsprint. In 1993, President Clinton required that the federal government purchase only paper containing at least 20% post-consumer-recycled paper by the end of 1994, and an increase to at least 30% by 1998.

The recycling of glass was increased by the Bottle Bill. The **Bottle Bill** requires consumers to pay a deposit on beverage containers at all types of food stores. After a beverage is consumed, the bottle is returned to the market and the deposit returned to the consumer. The decrease in containers going to landfills was significant.

Figure 7.24: WANTED! WHITE OFFICE PAPER. Businesses, government offices, and schools generate tons of white paper waste daily. Collecting and then using recycled paper reduces stress on our natural resource supply.

SUMMARY

All in all, we must be environmentally conscious of whatever we do. It takes clear thinking and the cooperation of all the Earth's inhabitants to conquer pollution. Local, state and national governments worldwide must join together with their citizens and fight to preserve our world. We strive as good stewards to leave the Earth for future generations in a condition better than we found it. With an attitude such as this worldwide, we would have no fear of our future generations gasping for clean air or wading in trash. They will enjoy the beauty of this creation that we have been given to enjoy.

Think about your home, school, church, neighborhood, or city. What could you do to help? What little step for man's benefit could you take to make a difference for your future offspring? Think up a small habit change that would lessen the need for new landfills and then make it a lifelong action on your part. Every little bit may hurt or help. Which end of the scale would you like to be weighing in on? "A good steward **is** as a good steward **does**!"

Figure 7.25: NEW SHAPE FOR GRASSES. Grasses, grapevines, and straw are used today in many craft projects. What other natural resources are "crafters" searching for and using?

Figure 7.26: GOOD STEWARDS. Volunteers are planting spartina, a salt-tolerant grass along a river bank; this helps reduce erosion. It's time that we all "dug" in.

↘ DID YOU KNOW?

- 1970—3% of aluminum cans were recycled; glass, 1%.
- 1993—62%, cans recycled; glass, 25%.

✔ THINGS TO KNOW AND DO:

Define:

*recycling ____

____

____

*The Bottle Bill ____

____

____

List 5 types of materials that can be recycled.

1. _______________________________   3. _______________________________

2. _______________________________   4. _______________________________

5. _______________________________

How has government affected recycling?

✎ IN SEARCH OF...

Pick 5 household products. Create a use for the left-over materials after the product has been used. Explain in detail (or by diagram, drawing) each of the 5 recycled items.

Glossary

abiotic factors all nonliving, physical factors in our environment: sunlight, water, soil, and atmosphere (includes weather).

abiotic variables the components that consist of the soil, sediments, organic and inorganic matter, and "litter" (decayed insects, plants, etc.).

acetylsalicylic acid (aspirin) made from salicylic acid from willow tree bark is used to reduce fever and pain.

acid rain any form of precipitation that has been polluted by acids. Nitric and sulfuric acids are commonly found in acid rain.

aeration a process that sprays water into the air removing the bad taste and odor.

aerobic bacteria agents of decay that work on wastes such as dead fish and break it down into nutrients, phosphates, and CO_2.

aerobic microbes microorganisms that use O_2 in their respiration process.

air pollution pollutants enter the air and change the colorless, odorless, combination of gases we call air into a grayish, foul-smelling, smog.

"Alpine" tundra exists on high mountain peaks.

ammonification the process by which ammonia (NH_3) is produced by decomposition.

annual incident radiation the light falling on a plant that is converted to chemical energy.

annuals plants that live for one growing season, produce seeds, and then die.

aquatic biome water biome. Includes both salt and fresh water.

"Arctic" tundra (See tundra).

arsonists people who purposely set fires.

atmosphere the area within the biosphere that consists of gases, airborne matter (dust), and water vapor surrounding the Earth.

autotrophs organisms that make their own food through photosynthesis and are commonly called self-feeders, such as green plants.

Azotobacter a nitrogen-fixing organism (bacterium) which lives in the soil.

benthic zone deep ocean zone where the pressure is 1,000 atmospheres (about 15,000 lbs. per square inch).

biogeochemical cycles cycles dealing with the movement of energy and nutrients taken in, used, and released time and time again.

biological rhythms the pattern or cycle in organisms such as the flowering of plants, yearly migrations, and breeding cycles.

biomass the total dry weight of all living matter in an ecosystem.

biomes groups of ecosystems that have similar types of vegetation.

biosphere the total region where life exists, including the terrestrial and subterrestrial environments as well as both fresh and marine waters.

biotic factors living organisms such as plants, fungi, animals, and man.

black lung a disease caused by inhaling coal dust from coal mines.

block cutting the harvesting of trees in large, block-like sections.

bottle bill law required consumers to pay a deposit on beverage containers at all types of food stores.

camouflage organisms that blend in with their environment.

canopy a covering that is created by the tallest trees in an area.

carbon (C) a naturally occuring element found in the remains of organisms that slowly change into natural gas, petroleum, and coal that is later tapped for use as fossil fuels (see also: **photosynthesis**).

carnivores meat-eating organisms.

carrying capacity represents the number of individuals an area can support in terms of food, shelter, and space.

catalytic converters an anti-pollution device in today's cars, removes most pollutants that automobile engines produce.

cellulose a plant cell wall material.

chaparral forest containing broad-leafed evergreen shrubs and a variety of small trees.

chaparral area of close growth of low, evergreen oaks; winters are mild and rainy; and the summers are long, hot, and dry.

chlorination water disinfection process that kills bacteria.

chlorofluorocarbons (CFC's) have the ability to thin the ozone layer.

chlorophyll a molecule in green plants used to make food.

chloroplast the structure within a cell that contains chlorophyll.

clay very fine particles of rock.

Clean Air Act in 1967 established the standard set for air pollution control.

climate weather conditions that have been recorded over a period of time in an area.

climax community a community that results from the replacing of organisms one after another.

CO2 a gas in the atmosphere used by green plants in the process of photosynthesis (see also **carbon**).

coagulants chemicals that form sticky globs called floc.

commensalism one organism benefits and the other is neither helped nor harmed in a symbiotic relationship.

communities different populations living within the same area.

competition the act of striving for food, water, shelter, mates, etc. by a number of organisms located in the same habitat.

conservation includes the wise use, the management, and protection of natural resources.

conservation tillage the number of times a field is plowed each year.

contour plowing land plowed back and forth across the slope rather than up and down.

controlled burning clearing out dense undergrowth by fires; keeps the forest "clean" and also minimizes the effect of destructive forest fires.

cover crops crops that keep the soil covered to prevent soil loss.

crop rotation rotating crops that are planted in a field to keep the soil fertility level high.

cytolysis a bursting of the cell.

dam a man-made structure to hold runoff water in times of heavy rainfall, preventing flooding and reducing soil erosion.

deciduous trees have broad-shaped leaves that are shed seasonally (oak, maple).

decomposers organisms that break down matter into inorganic substances.

deep well well drilled deep into the underground water supply.

dehydration the term used to describe the loss of water from the body.

denitrification a process by which nitrogen is returned to the atmosphere.

density independent factors abiotic factors such as air, water quality, temperature, soil, and wind rate.

desalination the removing of salt from seawater.

deserts areas of high temperature and low rainfall.

detention the temporary storage of water on land or in oceans.

detritus eaters organisms that feed on decaying material.

diatoms microscopic one-celled organisms found in the ocean.

digitalis a heart stimulant obtained from the leaves of the purple foxglove plant.

distillation heating and condensing water vapor to produce pure water.

dominant species the species in an ecosystem that has the largest population.

doubling rate the time it takes for a population to double.

dredging a scoop that digs up sand and gravel that contains minerals which then are deposited and washed.

ecological succession orderly replacing of various communities one after another in an area.

ecology the study of our environment, the interactions and relationships of all living and nonliving things.

ecosystem a community of plants and animals in their environment.

electrodialysis electric current passed through water, salt breaks into positive and negative ions; salt is then drawn off, leaving fresh water.

electromagnetic radiation consists of electromagnetic rays produced from x-ray machines, TV sets, microwave ovens, lasers and computers.

endangered wildlife seriously facing extinction.

Environmental Protection Agency (EPA) carries out the requirements of the environmental laws.

epiphytes aerial plants that live on the limbs and trunks of trees (orchids).

erosion the gradual wearing away of soil by rain and wind.

euphothic zone upper levels of aquatic ecosystems where light can penetrate.

eutrophication nutrient enrichment causing a dense growth of plant and animal life.

evaporation the release of water as vapor to the atmosphere.

evergreen forests forests that contain spruce, balsam fir, pine, cedar, hemlock and Douglas fir.

evergreen trees (conifers) have needle-like leaves not shed seasonally (pine, spruce).

extinction a dying out of a species.

fauna all the animal life in an area.

feces solid waste products of organisms.

flora all the plant life in an area.

fluorocarbons substances such as aerosol propellants, refrigerants, lubricants, solvents in making plastic, and automobile exhaust.

food a term that refers to material that contains chemical energy for sustaining life.

food chain the transfer of energy from a producer (green plants) to a consumer (insect, animal, and man).

food web a complex food chain.

fossil fuels nonrenewable resource such as oil, natural gas, coal, and gasoline that are used as an energy source. Fossil fuels come from the preserved remains of once living things.

freezing process produces pure water ice crystals from salt water; salt is separated and trapped from the fresh water ice crystals.

geothermal energy energy that is generated from the heat within the Earth.

global warming an increase in the temperature of the Earth due to an increase of CO_2 in the atmosphere.

glucose a simple sugar that is produced in photosynthesis.

grasslands area of transition between the temperate forests and the desert.

greenhouse effect the atmosphere absorbs heat from the sun and holds this heat for a time close to the Earth's surface. Carbon dioxide and water vapor trap this heat, increasing the temperature of the atmosphere.

ground water water found in pores of rock and sediment in fully saturated areas.

habitat special living space within a community.

"hard" water an abundance of minerals in water.

hardwood forests forests that contain maple, hickory and oak trees.

herbivoires those groups of organisms that directly feed on producers.

heterotrophs organisms which cannot make their own food (consumers).

homeostasis a steady state or balance.

hydrologic cycle is the movement of water throughout the biosphere.

hydrosphere the term used for the total of all water, frozen or liquid, on or near the Earth's surface.

improvement cutting diseased, crooked, or aged trees are removed; less valuable species are also removed allowing more room for the more commercially desirable ones to grow.

inexhaustible resources resources that cannot be used up.

landfills large open areas that contain most of our solid wastes.

Law of Conservation of Energy energy can neither be created nor destroyed but can be changed from one form to another.

laws of thermodynamics the first law states that energy is neither created or destroyed, but can be changed from one form to another. The second law states that with any energy conversion, some of the energy is lost to the system in the form of heat.

leeward refers to the side of the mountain away from the wind.

legumes plants such as peanuts, peas, and beans that are able to fix nitrogen.

lianas woody vines.

lichens organism composed of algae and fungi existing in a symbiotic relationship.

littoral zone intertidal zone along the shores of oceans.

marquis (See chaparral).

matorral (See chaparral).

mechanical action a process by which large pieces of matter are broken down into small pieces by the feeding process of certain organisms.

metabolism the sum total of all life processes.

microflora microscopic plant life that inhabits (lives on) plants.

minerals matter found in the Earth's crust and oceans such as iron, lead, copper, zinc, gold, silver and nickel.

mutualism two organisms living together, each benefitting from the relationship.

national forests forests are owned and operated by the federal government.

natural resources naturally occuring resources: minerals, air, soil, water, forests.

niche the particular role or lifestyle of a specific population in a habitat.

nitrification the process by which nitrogen is converted to nitrate ions (NO_3).

nitrogen (N_2) a gas that makes up about 78% of the Earth's atmosphere; it is also found in proteins, nucleic acids (RNA, DNA), and oceans.

nitrogen assimilation the process by which nitrogen is incorporated into organic molecules of living organisms (plants and animals).

nitrogen-fixing bacteria are bacteria living in the roots of special plants (legumes) that take nitrogen from the air, and through a series of reactions, "fixes" it into a more usable form for the plant's benefit.

noise pollution noise from airplanes, music, construction and industry that causes discomfort to many people and animals.

nonrenewable resources resources that can not be replaced.

open pit mining recovers minerals that are close to the Earth's surface (copper).

"open system" flowing of energy and nutrients into and out of the ecosystem.

ores elements found in the soil of the Earth's crust that are mined for their economic value.

oxygen (O_2) an element which is necessary for living things. Oxygen makes up about 21% of the Earth's atmosphere.

ozone layer layer of gas in our atmosphere that protects plants and animals from harmful ultraviolet light.

pampas (See grassland).

paramecium a single-celled organism.

parasitism a symbiotic relationship where one organism lives off of another organism (host) for much or all of its life; host is harmed in some way.

particulates tiny particles.

perennials plants that have more than one growing season.

permafrost a layer of subsoil permanently frozen.

photosynthesis the process by which the producer (green plants), in the presence of sunlight, makes food.
Green plant + sunlight + CO_2 + H_2O + Minerals —> Carbohydrates + O_2.

photosynthetic capacity the ability of a plant to achieve its capacity of food production in a given time.

phytoplankton microscopic organisms spread across the entire ocean surface.

pioneer organisms first organisms to settle in an area.

placer mining to separate, by washing, heavy metals such as gold, tin, and platinum from sand and gravel deposits.

plankton microscopic aquatic organisms.

plasmolysis the shrinking of the cell.

poaching taking wildlife that are endangerd or not in season.

pollutants change clean air into polluted air. Some air pollutants are carbon monoxide, smoke, nitrogen oxides, ozone, and sulfur dioxide.

potable water fit for drinking.

precipitation water that comes from the atmosphere to the earth as rain, sleet, snow, or hail.

predation organisms that hunt and kill prey.

prey the victims of carnivores.

primary consumers the second trophic level organisms that feed on the producers.

primary succession a potentially habitable ecosystem that is essentially lifeless and usually devoid of soil and available for colonization.

producers largely green plants that are autotrophs.

pyramid of numbers a pyramid that represents the number of individual organisms at each trophic level.

pyramid of biomass a pyramid that represents the information collected on the amount of biomass in an ecosystem.

quarrying a process that removes large chunks of matter that are found close to the surface level.

quinine a substance derived from the bark of cinchona tree; is used to treat malaria.

radiant energy (light energy) emitting heat or light (ex.: sunlight).

radon naturally occuring gas in the Earth's crust that may leak through basements and pipes.

rain shadow a reduction in rainfall amount on the leeward side of a mountain range.

rare wildlife has small populations but numbers are not decreasing.

rate regulators consumers that determine how fast nutrients are recycled in the ecosystem.

reclamation restoring the mined land to its original condition.

recyclable material may be reused over again.

reforestation a program that aids wildlife conservation by planting shrubs and trees providing cover for many forms of wildlife reducing the rate of soil erosion.

renewable resources organisms that can be used and replaced, such as plants and animals.

reverse osmosis high pressure causes fresh water to be squeezed through a membrane, leaving the salt behind.

Rhizobium nitrogen-fixing organisms that live in a symbiotic relationship in the roots of plants.

runoff water moving on the surface of the land.

Safe Drinking Water Act law passed to keep the nation's water supply safe.

salinity salt content of marine environment.

sand large particles of rock formed from the weathering process.

saprobes decomposers which use dead matter for nutrition.

saprophytes organisms that live on dead matter and cause decomposition of that matter.

savanna a tropical grassland that is covered with grasses, shrubs, and a few scattered trees.

secondary consumers organisms such as foxes, owls, whales, and spiders that feed on primary consumers.

secondary succession succession in an area following destruction of an existing community by fire or storm.

senescence the term used to describe old age in plants.

sewage treatment plants plant and animal matter is removed by the use of bacteria and oxygen in breaking down organic matter and turning it into inorganic nutrients.

shallow well well drilled into the water table.

shelter belts trees or large thick shrubs are planted in wide rows to break or slow down the wind speed.

silt medium-sized particles of rock.

smog thick layer of polluted air lying near highly industrialized cities.

sodium chloride (NaCl), table salt.

soil a material that formed by wathering of rock.

solid waste litter along roadsides, on beaches, parks, floating in oceans, ponds and lakes or heaped up in hughe mountains in dumps or in landfill.

species diversity varying of species from ecosystem to ecosystem.

stable an overall, unchanging.

stable dynamic equilibrium a fixed, moving balance within an environment.

steppes (See grassland).

strip cropping strips of row crops are alternated on slopes.

strip mining miners cut a furrow, placing the removed top layer in a ridge at the sides of the furrow; explosives break up and remove the ore.

subsoil made up of mainly rock particles but does contain tiny amounts of organic matter.

succession one organism replacing another in a specific environment over a period of time.

succulent desert plants adapted for water storage, storing water in their roots and stems.

symbiosis organisms living together in a specific relationship.

taiga northern coniferous forests.

temperate forests forests located north and south of the tropics in areas called temperate zones. Climate here is mild, but may range from below freezing in some months to a sizzling 100° F in other months. Rainfall also varies in temperate areas from dry to moderately wet.

terracing wide, flat rows that are built along hillsides; they hold water, preventing it from washing soil away down a hillside and forming gullies.

tertiary consumers organisms that feed on the secondary consumers.

thermal inversion a layer of warm air settles above a layer of colder air and holds this cold air mass close to the ground, an occurance due to increased air pollution levels.

thermal pollution occurs when heated water is dumped into ponds, lakes, and rivers.

threatened wildlife may be abundant in some areas but they are still facing serious dangers.

topsoil the upper level of soil containing organic matter (humus), rock particles, and many tiny organisms.

transition zones areas between distinct biomes where characteristics gradually end, changing old characteristics for new ones.

transpiration the process by which water vapor is released through tiny openings (stomata) on the underside of leaves.

transportation water moving from one place to another.

treeline the highest point where trees can grow.

trophic levels hierarchy of energy transfers in an ecosystem.

tropical rainforest ecosystem having direct sunlight, warm temperatures, and daily rainfall.

tundra northern most area, where summer temperatures average between 0° to 45°C and winter temperatures range between -34° to -40°C. Summers are short and mild; winters long and cold.

U.S. Fish and Wildlife Service a government agency that strives to conserve plants, soil, and water.

ultraviolet (UV) rays rays of the sun just beyond the violet in the visible spectrum on the border of the x-ray region.

veld (See grassland).

water pollution polluting substances in water from industries, farms, and sewage treatment systems.

water table water deposited close to the surface of the land.

watershed an area where rain falling on the land is funneled into a single stream that runs down the slopes of a valley.

watershed management another way of increasing water supply for man by reducing runoff of rainwater and melting snow.

weathering the process that breaks down rock material into smaller particles from which soil is formed.

weathering changing rock by wind, rain, glaciers, melting snow and plants into soil particles.

wetland very wet areas where the vegetation is partially or completely submerged.

white lung disease caused by inhaling dust from glass manufacturing.

wildlife includes all untamed animals and uncultivated plant life.

windward side is the side that takes on the wind directly.

zero population growth refers to the death rate and the birth rate overall being about equal.

zooplankton microscopic water organisms.

Bibliography

Arms, Karen. *Environmental Science.* Second Edition. Fort Worth, Philadelphia, San Diego, et al.: Saunders College Publishing, 1994.

Biological Sciences Curriculum Study. *High School Biology.* Second edition. Chicago: Rand McNally and Company, 1968.

Chadwick, Douglas H. *Enduring America.* Washington, DC: National Geographic Society, 1995.

Curtis, Helena. *Biology.* New York, New York: Worth Publishers, 1983.

Emmel, Thomas C. *An Introduction to Ecology and Population Biology.* New York: W. W. Norton and Company, Inc., 1973.

Graham, Edward H., and William R. VanDersal. *Water for America—The Story of Water Conservation.* New York: Henry Z. Walck, Inc., 1956.

Graham, Edward H., and William R. VanDersal. *Wildlife for America—The Story of Wildlife Conservation.* New York: Henry Z. Walck, Inc., 1949.

Haskel, Sebastian, and David Sygoda. *Fundamental Concepts of Modern Biology.* New York, NY: Amsco School Publications, Inc., 1972.

Heimler, Charles H. *Ecology and Human Influence.* Columbus, Ohio: Charles E. Merrill Publishing Co., 1977.

Kaufman, Donald G., and Cecilia M. Franz. *Biosphere 2000—Protecting Our Global Environment.* Dubuque, Iowa: Kendall/Hunt Publishing Company, 1996.

Kraus, David. *Concepts in Modern Biology.* 5th ed. Englewood Cliffs, New Jersey: Globe Book Company, 1984.

Lerman, Matthew. *Marine Biology—Environment, Diversity, and Ecology.* Menlo Park, California: The Benjamin/Cummings Publishing Company, Inc., 1986.

McConnaughey, Bayard H., Ph.D. *Introduction to Marine Biology.* Second edition. Saint Louis: The C. V. Mosby Company, 1974.

National Geographic Society. *Animal Kingdoms—Wildlife Sanctuaries of the World.* Washington, DC: National Geographic Society, 1995.

National Geographic Society. *Our Inviting Eastern Parklands—from Acadia to the Everglades.* Washington, DC: National Geographic Society, 1994.

Reader's Digest. *Explore America—National Parks.* Pleasantville, New York/Montreal: The Reader's Digest Association, Inc., 1993.

Ritchie, Donald D., and Robert Carola. *Biology.* Second edition. Reading, Massachusetts and Menlo Park, California, et al.: Addison-Wesley Publishing Company, 1979.

Slesnick, I., L. Balzer, A. McCormack, D. Newton, and F. Rasmussen. *Biology.* Glenview, Illinois: Scott Foresman, 1980.

Smith, Robert Leo. *Elements of Ecology.* Second edition. New York: Harper and Row Publishers, 1986.

Starr, Cecie, and Ralph Taggart. *Biology—The Unity and Diversity of Life.* Belmont, California: Wadsworth Publishing Company, 1981.

Starr, Cecie. *Biology—Concepts and Applications.* Belmont, California: Wadsworth Publishing Company, 1991.

Wallace, R., G. Sanders, and R. Ferl. *Biology—The Science of Life.* NY, NY: Harper Collins Publishers, 1991.

Index